Another Fine Math
You've Got Me Into...

Another Fine Math Youve Got Me Into...

Ian Stewart

W. H. Freeman and Company
NEW YORK

Library of Congress Cataloging-in-Publication Data

Stewart, Ian.
    Another fine math you've got me into- / Ian Stewart.
        p.   -cm.
    Includes bibliographical references.
    ISBN 0-7167-2342-5. — ISBN 0-7167-2341-7 (pbk.)
    1. Mathematical recreations.   I. Title.
QA95.S723   1992
793.7′4 — dc20   92-13455
                                                                     CIP

PRINTED IN THE UNITED STATES OF AMERICA

1 2 3 4 5 6 7 8 9 0 VB 9 9 8 7 6 5 4 3 2

Contents

	Foreword by Martin Gardner	vii
	Preface	ix
1	The Lion, the Llama, and the Lettuce	1
2	Tile and Error	19
3	Through the Evolvoscope	33
4	Cogwheels of the Mind	51
5	How Many Goats in the Orchard?	65
6	Passage to Pentagonia	79
7	Knights of the Flat Torus	93
8	A Vine Math You've Got Me Into	113
9	Maxdoch Murwell, Market Manipulator	131
10	Curie's Mistake	145
11	A Dicey Business	161
12	The Thermodynamics of Curlicues	177
13	The Group-Theorist of Notre Dame	199
14	A Six-Pack for the Tree-God	221
15	The Well-Tempered Calculator	235
16	Sofa, So Good . . .	255

Foreword

I first encountered Ian Stewart two decades ago when I subscribed to *Manifold*, a whimsical journal compiled quarterly by students at the Mathematical Institute of the University of Warwick, in Coventry, England. Stewart contributed frequently to this outrageous magazine. I particularly recall his dialogue between Rosen Cranz and Guilden Stern on the inverted Goldbachian conjecture that every prime is the sum of two even numbers.

Stewart later coedited *Seven Years of Manifold* (Shiva Publishing, 1981), the best collection of mathematical humor ever assembled. Its frontispiece depicted "the mating dance of the Alexander Horned Spheres." Inside were such gems as a picture of Bertrand Russell shaving all those who don't shave themselves, a necessarily short paper on the one-color map theorem, notes on how to knit a Klein bottle, and the riddle: What's purple and commutes? (Answer: an Abelian grape.)

Stewart soon became an editor of *The Mathematical Intelligencer* and began writing his marvelous articles and books. He is that rare combination of one who knows and loves mathematics on all its bewildering levels and can write about them with zest and humor, in ways anyone can comprehend. To these skills he adds a fondness for recreational topics and amusing wordplay. It is impossible to read him without learning a great deal and thoroughly enjoying the instruction.

Martin Gardner

Preface

During my last few years at school one of the high points of my existence was the arrival of *Scientific American* and Martin Gardner's "Mathematical Games" column. Now, thirty years on, I find myself as the fourth occupant of the position that he created. It's an odd feeling, though a kind of "anthropic principle" explains the coincidence. Anyone even remotely suitable to follow in Gardner's footsteps is *bound*, as a teenager, to have had the type of mind that would have been attracted to his column.

It still feels strange.

I'm going to explain how I inherited his mantle; partly because it exemplifies a favorite theme of mine, that nothing ever happens the way you'd expect it to, and partly because it explains the book that you now hold in your hands. It is *not* a collection of *Scientific American* columns —though those that I now write may well, fate willing, see daylight in a similar format. It is a selection of columns from *Pour la Science*, the French edition of that magazine, many of which also appeared in other European editions.

There are many national editions of *Scientific American.* They don't just translate the American edition word for word; they include their own articles, and a lot of the editorial matter is different, because different countries have different concerns. Gardner's column passed on to Douglas Hofstadter as "Metamagical Themas," and then bowed to

changing interests and became "Computer Recreations," written by
A. K. Dewdney. Eagle-eyed readers may have noticed that the title
recently changed to "Mathematical Recreations," reflecting a trend
back toward its original subject matter.

When "Computer Recreations" first started up, Philippe Boulanger,
the editor of the French edition, thought it was a great idea. However,
he also wanted to retain the interest of people who liked mathematical
games but weren't computer freaks. So *Pour la Science* translated
"Computer Recreations," but it also started up its own column of "Jeux
Mathématiques," with a succession of authors. Eventually I started
writing that column, and it became "Visions Mathématiques": the column remained recreational but began to include material less overtly
related to games.

How did an Englishman end up writing a regular column in a French
magazine? It began with a French physicist, Jean-Pierre Petit. He had
produced some informal notes for his students in comic-book format, on
aerodynamics, black holes, and so forth. Some of them were published
by the Librairie Classique Eugène Belin, the publishers of *Pour la
Science*. They were enormously successful, and at the instigation of
Christopher Zeeman—then my boss and a friend of Petit's—I was
signed up to translate some into English. I was a scientifically literate
amateur cartoonist with a quirky sense of humor, and it was thought that
I might be sympathetic to what Petit was trying to achieve.

Philippe then got the idea that I ought to produce some mathematical comic books for the same series. I would write them in English, he
would translate them into French, and everybody would pretend that the
French version was the original and the English version was a translation. That worked very well, and some people from the satirical magazine *Le Canard Enchaîné* were brought in to add some French jokes.
Round about this time the most regular contributor to "Jeux Mathématiques" decided that he no longer had the time to write the column, and
Philippe approached me, asking whether I knew anyone who could take
over on a semiregular basis.

I sure did.

A couple of years ago, several other foreign-language editions of
Scientific American agreed to run the column as well. Then the parent
magazine in the States decided to let me share its "Mathematical Recreations" column with Dewdney. Now I write six columns per year, which

alternate with "The Amateur Scientist." I also write six extra columns for the French edition, so that it is monthly in France (and Spain) but bimonthly in America, Britain, and in all the other national editions.

It gets confusing sometimes.

This book, then, is a selection of sixteen columns from those written for *Pour la Science*. The style is identical to the columns you may have read in *Scientific American*. A further twelve columns have already been published under the title *Game, Set, and Math*. The material has been edited a little to take advantage of any new developments since they were first written.

I make no claims about emulating Martin Gardner's style. It can't be done: Gardner is unique. My style has settled into a kind of fictional narrative, in which weird characters such as the Worm family (Henry, Anne-Lida, and baby Wermentrude), Maxdoch Murwell the international financier, and Snitchswisher Wishsnitchersdorter the neolithic numero-sophist undergo close encounters of a mathematical kind. The stories are fun seasoned with a pinch of seriousness. Most of them are based upon some significant mathematical idea, which I hope emerges from the byplay and sticks in readers' minds.

For instance, "The Lion, the Llama, and the Lettuce" tells how Weffolk farmer Algernon Quinn got his produce to market; it's also about graph theory. "Through the Evolvoscope" talks of flying cats and flipperpotami — and also about catastrophe theory. It's clear what area of mathematics "The Group-Theorist of Notre Dame" must be about, but you don't often meet the hunchback and Tarzan's beloved in the same story.

Games and recreational mathematics remain paramount. In "Tile and Error," Henry Worm sets out to tile his bathroom and succeeds only with the help of Albert Wormstein. Merlin's efforts to amuse his king in "Knights of the Flat Torus" end with one of the worst puns in the entire book. And Bumps the goose-girl and Grimes the shepherd-boy discover surprising strategies in offbeat dice games.

I had enormous fun writing all this stuff. I hope some of the enjoyment rubs off on you.

Ian Stewart

●

The Lion,

●

the Llama,

●

and the Lettuce

●

─────

●

Along one of the dusty country roads that are typical of the county of Weffolk came a farmer. In his right hand he clutched a gigantic lettuce. His left hand grasped two rope halters. Ahead, on one halter, ambled a llama. Behind, on the other, prowled a lion. A curious procession, you may think, but such sights are commonplace in the rugged Weffolk countryside, a region noted — especially on Fridays — for its idiosyncratic agriculture. Friday is market day, and Algernon Quinn was taking his produce to market.

And Quinn had a problem. The bridge across Rising Gorge had collapsed and had been temporarily replaced by a breeches buoy. It was strong enough to carry only Quinn, together with one item of produce — lion, llama, or lettuce. (As I said, it was a *gigantic* lettuce. And a fairly hefty llama, if the truth be known.)

Trivial, I hear your scathing dismissal. *Take the lion across, return for the llama, and finally transport the lettuce.* Obviously you're no farmer. Any true son of the soil knows, intuitively and without logical thought, what such a plan will lead to. On returning from transporting the lion, Al Quinn will find a fat and happy llama, but no lettuce. A llama will guzzle an entire lettuce, however gigantic, at a single sitting. Indeed the plan has a second fatal flaw, for when a lion is left alone with a llama it tends to see the creature more as llamaburger. On the other hand, you don't expect to see a hungry lion prowling through the vegetable patch in search of a fat, juicy lettuce, so the vegetable may safely be left with the carnivore.

By now you have recognized Algernon Quinn's dilemma as the hoary wolf–goat–cabbage puzzle in disguise; and perhaps you also observed that Al Quinn is the reincarnation of the medieval mathematician Alcuin (735–804), to whom that puzzle is usually attributed. It is certainly quite ancient and appears in Ozanam's *Récréations Mathématiques et Physiques* of 1694. In one respect at least you are correct, for Algernon Quinn has the precise logical mind of a born mathematician. Not for him the trial-and-error approach, but systematic reasoning only. And Al reasons as follows:

"First I must simplify the problem and find its essential features. The important thing is which side of the gorge each of my three marketable items is on. It's irrelevant where *I* am, or where the breeches buoy is, because those are free to move at will. Subject only to the aforementioned gastronomical constraints, that the lion should not be left alone with the llama, nor the llama with the lettuce.

"I can represent the position of a single item by the digits 0 and 1, using 0 to represent this side of the gorge and 1 to represent the far side. Thus the configuration of all three items is represented by a triple (L, λ, l) in three-dimensional lion–llama–lettuce space. For example $(L, \lambda, l) = (1, 0, 1)$ represents $L = 1$, $\lambda = 0$, $l = 1$; that is, the lion on the far side, the llama on this side, and the lettuce on the far side.

"How many configurations are there? Well, each coordinate L, λ, or l can take one of the two values 0 or 1. Thus there are $2 \times 2 \times 2 = 8$ possibilities. What's more, they have a beautiful geometric structure: they are the eight vertices of a unit cube in lion–llama–lettuce space (Figure 1A).

"I may move only a single item at a time; that is, I may traverse only the edges of the cube. But some edges are forbidden. For example, the edge from $(0,0,0)$ to $(1,0,0)$ corresponds to taking the lion across the gorge on its own. But this leaves llama and lettuce unchaperoned, so I would shortly be greeted by a fat llama and no lettuce. In fact these gastronomic constraints rule out exactly *four* edges, which I will draw as dashed lines. The rest, representing permissible moves, I will make solid.

"The problem thus geometrized becomes: can I start at (0,0,0)—all items on *this* side—and get to (1,1,1)—all items on the *other* side— passing only along edges of the cube that are solid lines? And of course the answer is 'yes.' Indeed, from a topological viewpoint, I can lay the

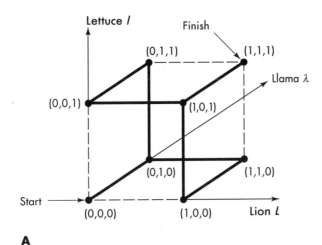

A

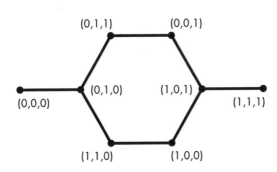

B

FIGURE 1   *A.* Possible moves graphed in lion–llama–lettuce space. Dashed edges are forbidden, solid edges are permitted. *B.* A simplified graph of the solid edges makes the two distinct answers obvious.

edges out flat (Figure 1*B*) and the solution stares me in the face. *Two* solutions, in fact, and *only* two if I avoid unnecessary repetitions [see the box below]. They differ only by a lion/lettuce symmetry operation."

Al Quinn's geometric method applies to a huge range of puzzles, in which objects must be rearranged according to certain rules, and the object is to get from some given starting position to some given finishing position. The idea is to form a *graph* consisting of vertices (dots) joined by edges (lines). Each vertex corresponds to a position in the puzzle, and each edge corresponds to a legal move. The solution of the puzzle is then a path through the graph, joining the starting vertex to the finishing one. Such a path is usually obvious to the eye — provided the puzzle is

HOW TO GET ACROSS WITH YOUR PRODUCE INTACT

SOLUTION 1

(0,0,0)	Start	
(0,1,0)	Take llama over	
(0,1,1)	(Return and) take lettuce over	
(0,0,1)	Bring back llama	
(1,0,1)	Take lion over	
(1,1,1)	(Return and) take llama over.	

SOLUTION 2

(0,0,0)	Start
(0,1,0)	Take llama over
(1,1,0)	(Return and) take lion over
(1,0,0)	Bring back llama
(1,0,1)	Take lettuce over
(1,1,1)	(Return and) take llama over.

sufficiently simple that the entire graph can be drawn. Puzzles of this type are really disguised mazes, for a maze is just a graph drawn in a slightly different fashion.

PROBLEM ❶

The following week, Al Quinn took to market a lettuce, a llama, a lion, and a leviathan. The bridge was still down. Unsupervised leviathans, as you know, eat lions —unless a lettuce is also present, for leviathans become docile when subjected to the smell of fresh lettuce. Draw up the graph (it may or may not be helpful to observe that it is a unit hypercube in leviathan-lion-llama-lettuce space, with coordinates ($\mathcal{L},L,\lambda,l$) all 0 or 1, and some edges deleted) and see whether or not a solution exists.

Although Quinn's graphical approach is applicable in principle to many puzzles, there is often a practical snag: if the number of positions or moves is too large, then the graph cannot be drawn. For example, in principle the Rubik cube could be solved by drawing its graph—but the graph would need 43,252,003,274,489,856,000 vertices! The next problem is somewhere toward the limits of what is possible in practice, and also illustrates that a bit of extra thought may lead to a simpler solution.

PROBLEM ❷

Use Al Quinn's graphical method to find a way to slide the three blocks (Figure 2), without turning them, so that they all occupy the right-hand side of the region shown.
Does the block puzzle remind you of anything simpler? How does this help?

Another traditional puzzle leads to a graph of considerable beauty. The Tower of Hanoi was marketed in 1883 by the great French recreational mathematician Edouard Lucas (under the pseudonym M. Claus). In 1884, in *La Nature*, M. De Parville described it in romantic terms:

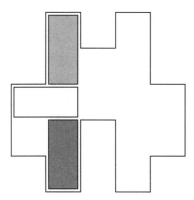

FIGURE 2   Can you slide the blocks to the right-hand side of the region?

In the great temple at Benares, beneath the dome which marks the
centre of the world, rests a brass plate in which are fixed three
diamond needles, each a cubit high and as thick as the body of a bee.
On one of these needles, at the creation, God placed sixty-four discs of
pure gold, the largest disc resting on the brass plate, and the others
getting smaller and smaller up to the top one. This is the Tower of
Bramah. Day and night unceasingly the priests transfer the discs from
one diamond needle to another according to the fixed and immutable
laws of Bramah, which require that the priest on duty must not move
more than one disc at a time and that he must place this disc on a
needle so that there is no smaller disc below it. When the sixty-four
discs shall have been thus transferred from the needle on which at the
creation God placed them to one of the other needles, tower, temple,
and Brahmins alike will crumble into dust, and with a thunderclap the
world will vanish.

The Tower of Hanoi is similar to the Tower of Brahma but with eight (or
sometimes fewer) discs. It is an old friend of recreational mathematicians, and it may seem that little new can be said about it. But, as we
shall see, Al Quinn's graphical approach leads to a delightful surprise,
fully in tune with the modern era.

For definiteness, consider 3-disc Hanoi, that is, the Tower of Hanoi
with just three discs. Sample positions and legal moves are shown in
Figure 3. To construct the graph, we must first find a way to represent
all possible positions, then work out the legal moves between them, and

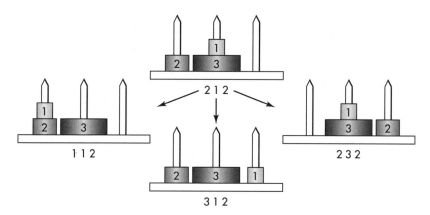

FIGURE 3   The three legal moves from position 212.

finally draw up the graph. I'll describe what I actually did, because to begin with it's not obvious how to proceed — and then we'll observe, with twenty-twenty hindsight, that there is a much cleverer method.

How can we represent a position? Number the three discs 1,2,3, with 1 being the smallest and 3 the largest. Number the needles 1,2,3 from left to right. Suppose that we know on which of the three needles each disc is: for example, disc 1 is on needle 2, disc 2 on needle 1, and disc 3 on needle 2. Then we have completely determined the position, because the rules imply that disc 3 must be *underneath* disc 1. We can encode this information in the sequence 212, the three digits in turn representing the needles for discs 1, 2, and 3. Therefore each position in 3-disc Hanoi corresponds to a sequence of three digits, each being 1, 2, or 3. To make this clear, Figure 3 includes these codes.

It follows that there are precisely $3 \times 3 \times 3 = 27$ different positions in 3-disc Hanoi. But what are the permitted moves?

The smallest disc on a given needle must be at the top. It thus corresponds to the *first* appearance of the number of that needle in the sequence. If we move that disc, we must move it to the top of the pile on some other needle, that is, we change the number so that it becomes the *first* appearance of some other number.

For example, in the position 212 above, suppose we wish to move disc 1. This is on needle 2, and corresponds to the first occurrence of 2 in the sequence. Suppose we change this first 2 to 1. Then this is (trivially!) the first occurrence of the digit 1; so the move from 212 to 112 is legal. So is 212 to 312 because the first occurrence of 3 is in the first place in the sequence.

We may also move disc 2, because the first occurrence of the symbol 1 is in the second place in the sequence. But we cannot change it to 2,

THE LEGAL MOVES IN 3-DISC HANOI

START HERE . . .	FINISH ON ANY OF THESE		
111	211	311	
112	212	312	113
113	213	313	112
121	221	321	131
122	222	322	132
123	223	323	133
131	231	331	121
132	232	332	122
133	233	333	123
211	111	311	231
212	112	312	232
213	113	313	233
221	121	321	223
222	122	322	
223	123	323	221
231	131	331	211
232	132	232	212
233	133	333	213
311	111	211	321
312	112	212	322
313	113	213	323
321	121	221	311
322	122	222	312
323	123	223	313
331	131	231	332
332	132	232	331
333	133	233	

because 2 already appears earlier, in the first place. A change to 3 is, however, legal. So we may change 212 to 232 (but *not* to 222).

Finally, disc 3 cannot be moved, because the third digit in the sequence is a 2, and this is *not* the first occurrence of a 2.

To sum up: from position 212 we can make legal moves to 112, 312, and 232, and *only* these.

We can list all 27 positions and all possible moves by following the above rules; the result is shown in the box on page 8.

PROBLEM ❸

All but three positions give exactly three legal moves, but the other three positions give only two legal moves. Why?

The next task requires care and accuracy, but little thought. Draw 27 dots on a piece of paper, label them with the 27 positions, and draw lines to represent the legal moves. My first attempt at this ground to a halt in a mess of spaghetti. But after a bit of thought, rearranging the vertices and edges to avoid overlaps led to Figure 4.

Something that pretty can't be coincidence!

But before we investigate *why* the graph has such a regular form, let's observe that it answers the original question. To transfer all three discs from needle 1 (position 111) to needle 2 (position 222) we merely run down the left-hand edge, making the moves

$$111 \rightarrow 211 \rightarrow 231 \rightarrow 331 \rightarrow 332 \rightarrow 132 \rightarrow 122 \rightarrow 222.$$

Indeed, by consulting the graph it is clear that we can get from any position to any other — and it is also clear what the quickest route is.

PROBLEM ❹

A. What is the quickest route from 211 to 212?
B. What is the quickest route from 211 to 213?

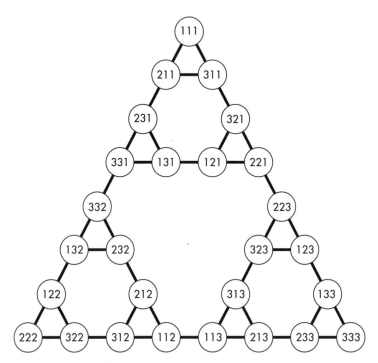

FIGURE 4   The graph for 3-disc Hanoi has a highly elegant form. Why?

On to a deeper question: what is the explanation for the remarkable structure of Figure 4?

The graph consists of three copies of a smaller graph, linked by three single edges to form a triangle. But each smaller graph in turn has a similar triple structure. Why does everything appear in threes, and why are the pieces linked in this manner?

If you work out the graph for 2-disc Hanoi you will find that it looks exactly like the top third of Figure 4. Even the labels on the vertices are the same, except that the final 1 is deleted. In fact it is easy to see this, *without* working out the graph again. You can play 2-disc Hanoi with three discs: just ignore disc 3. Suppose disc 3 stays on needle 1. Then we are playing 3-disc Hanoi but restricting our attention to those 3-digit sequences that end in 1, such as 131 or 221. But these are precisely the sequences in the top third of the figure. Similarly, 3-disc Hanoi with disc 3 fixed on needle 2 — that is, disguised 2-disc Hanoi — corresponds to the lower left third, and 3-disc Hanoi with disc 3 fixed on needle 3 corresponds to the lower right third.

This explains why we see three copies of the 2-disc Hanoi graph in the 3-disc graph. And a little further thought shows that these three subgraphs are joined by just three single edges in the full puzzle. To join up the subgraphs, we must *move* disc 3. When can we do this? Only when one needle is empty, one contains disc 3, and the other contains all the remaining discs! Then we can move disc 3 to the empty needle, creating an empty needle (the one it came from), and leaving the other discs untouched. There are six such positions, and the possible moves join them in pairs.

The same argument works for any number of discs. The graph for 4-disc Hanoi, for example, consists of three copies of the 3-disc graph, linked at the corners like a triangle. Each subgraph describes 4-disc Hanoi with disc 4 fixed on one of the three needles; but such a game is just 3-disc Hanoi in disguise. And so on (Figure 5). We say that the Tower of Hanoi puzzle has a recursive structure; the solution to ($n + 1$)-disc Hanoi is determined by that for n-disc Hanoi according to a fixed rule. The recursive structure explains why the graph for ($n + 1$)-disc Hanoi can be built from that for n-disc Hanoi. The triangular symmetry arises because the rules treat needles 1, 2, and 3 in exactly the same way. You can deduce the graph for 64-disc Bramah, or for any other number of discs, by repeatedly applying this rule to the graph for 0-disc Hanoi—which is a single dot! For example, Figure 6 shows the

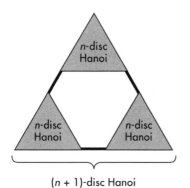

($n + 1$)-disc Hanoi

FIGURE 5   Recursive structure of n-disc Hanoi.

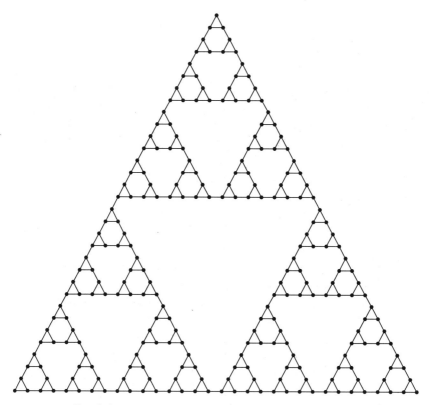

FIGURE 6 Graph for 5-disc Hanoi resembles a fractal known as the Sierpiński gasket.

graph for 5-disc Hanoi, drawn by applying this recursive structure. Brains instead of brawn! It would take hours to work 5-disc Hanoi by listing all 243 possible positions and finding all the moves between them—and you'd probably make several mistakes along the way.

PROBLEM ⑤

What is the minimum number of moves needed to move all n *discs from one needle to another in* n-*disc Hanoi?*

PROBLEM 6

How many distinct moves (i.e., edges of the graph) are there in n-*disc Hanoi?*

PROBLEM 7

I haven't labeled the vertices in Figure 6. No, I don't want you to do it! But I'd like you to show that it can be done in principle. Work out a rule for labeling the graph for (n + 1)-*disc Hanoi on the assumption that you know how to label the one for* n-*disc Hanoi.*

A final observation. As the number of discs becomes larger and larger, the graph becomes more and more intricate, looking more and more like the Sierpiński gasket.* This shape is a fractal, a geometrical object having detailed structure on all scales. This is a striking and surprising result, because the puzzle was invented almost a century before fractals were discovered. It is yet another demonstration of the remarkable unity of mathematics. Moreover, it has a curious application. Not long after this chapter's original appearance in the August 1989 issue of *Pour la Science,* I went to the International Congress of Mathematicians in Kyoto, and a German mathematician named Andreas Hinz introduced himself. He had been trying to calculate the average distance between two points in a Sierpiński gasket of unit side. One expert he had asked said it was "very difficult." Another said it was "trivial, and equal to $\frac{8}{15}$," but on closer analysis the proof didn't hold up. Hinz had already found a formula for the average number of moves between positions in the Tower of Hanoi puzzle.

In fact Hinz, and independently Chan Hat-Tung, found an exact formula for the average number of moves between positions in *n*-disc Hanoi. The *total* number of moves (using shortest paths) between all possible pairs of positions is given by the astonishing formula

---

*See *Game, Set, and Math,* Chapter 9.

$$\frac{466}{885}18^n - \frac{1}{3}9^n - \frac{3}{5}3^n + \left(\frac{12}{29} + \frac{18}{1003}\sqrt{17}\right)\left(\frac{1}{2}\left(5 + \sqrt{17}\right)\right)^n +$$
$$\left(\frac{12}{29} - \frac{18}{1003}\sqrt{17}\right)\left(\frac{1}{2}\left(5 - \sqrt{17}\right)\right)^n$$

which I exhibit as an example of the kind of thing that mathematicians can come up with.

Hinz and Chan didn't realize there was any connection with the Sierpiński gasket. Having read my article, Hinz saw that he could use his calculation for the Tower of Hanoi. Because there are 3^n positions, the average distance between two positions is asymptotic to $\frac{466}{885}2^n$, a value obtained by ignoring all except the first (largest) term in the formula and dividing by 3^n. That means that the ratio of the exact answer and this approximation tends to 1 as n becomes very large. Now the length of the side of the graph is 2^n, and dividing by that to make the side equal to 1 we get the answer $\frac{466}{885}$ in the limit of infinitely many discs. But the graph for Hanoi with infinitely many discs is the gasket. Therefore the average distance between two points in a unit Sierpiński gasket is $\frac{466}{885}$ *precisely*.

This is some 2% smaller than the value suggested by the second expert. Who says recreational mathematics has no serious payoff? For the statistically minded, Hinz also proved that the variance of the distance is precisely $\frac{904,808,318}{14,448,151,575}$. Any of you who keep an eye open for curious numbers in mathematics, I recommend that you add these two to your collection.

ANSWERS

1. The graph is now a unit hypercube in leviathan – lion – llama – lettuce space, with various edges deleted, as in Figure 7. One possible solution is as follows:

   Take llama across

   (Return and) take lion across

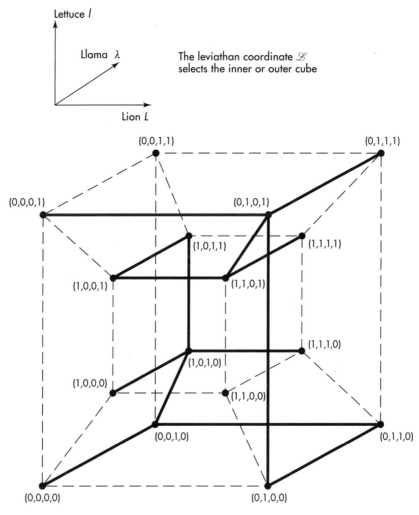

FIGURE 7 Hypercubic graph for the leviathan–lion–llama–lettuce puzzle. Dashed edges are forbidden, solid edges are permissible.

Bring back llama

Take lettuce across

(Return and) take leviathan across

Take llama across

There is another six-move solution: can you find it?

2. The graph is fairly large, with many loops and dead ends. You'll need a bigger sheet of paper than this page, and there isn't room to give the answer. But in any case, there's a quicker way to solve the puzzle, because it's a disguised version of lion–llama–lettuce. The darkest block is the llama.

3. In general each needle will have a different-sized disc on top. Then there are three possible moves: the small disc goes on either of the others, or the middle-sized disc goes on the largest. If one needle is empty, there are also three moves: place either top disc on the empty needle, or put the smaller disc on top of the larger. But when two needles are empty (and all the discs are in a single pile), there are only two moves: place the top disc on either empty needle.

4. $211 \rightarrow 231 \rightarrow 331 \rightarrow 332 \rightarrow 232 \rightarrow 212$
   $211 \rightarrow 311 \rightarrow 321 \rightarrow 221 \rightarrow 223 \rightarrow 323 \rightarrow 313 \rightarrow 213$

5. The number of *vertices* along a side of the graph doubles at each stage, and is thus 2^n for n-disc Hanoi. We want the number of *edges* along a side, which is one less than this, namely $2^n - 1$.

6. Suppose there are E_n edges in the n-disc graph. The recursive structure implies that $E_n + 1 = 3E_n + 3$; further, $E_1 = 3$. Thus $E_n = 3 + 3^2 + 3^3 + \ldots + 3^n = (3^{n+1} - 3)/2$.

7. Label the top subgraph for $(n + 1) -$ disc Hanoi exactly as for n-disc Hanoi, but with an extra 1 at the end. To get the lower left subgraph, turn the top subgraph (and the corresponding labels) counterclockwise 120°, and change 1 to 2, 2 to 3, and 3 to 1. To get the lower right subgraph, rotate clockwise 120° and change 1 to 3, 3 to 2, and 2 to 1.

FURTHER READING

Ball, W. W. Rouse. *Mathematical Recreations and Essays.* London: Macmillan, 1959.

Chan Hat-Tung. A statistical analysis of the towers of Hanoi problem. *International Journal of Computer Mathematics* 28 (1989): 57–65.

Graham, Ronald L.; Donald E. Knuth; and Oren Patashnik. *Concrete Mathematics*. Reading, Mass: Addison-Wesley, 1989.

Hinz, Andreas M. The tower of Hanoi. *L'Enseignement Mathématique* 35 (1989): 289–321.

———. Shortest path between regular states of the tower of Hanoi. *Information Science Abstracts*. Forthcoming.

———. The average distance on the Sierpiński gasket. *Probability Theory and Related Fields*. Forthcoming.

Kasner, Edward, and James R. Newman. *Mathematics and the Imagination*. London: Bell, 1961.

O'Beirne, T. H. *Puzzles and Paradoxes*. Oxford: Oxford University Press, 1965.

Ozanam, A. F. *Récréations Mathématiques et Physiques*. Paris, 1694.

Rubik, Ernö; Tamás Varga; Gerszon Kéri; György Marx; and Tamás Verkedy. *Rubik's Cubic Compendium*. Oxford and New York: Oxford University Press, 1986.

Stewart, Ian. *Game, Set, and Math*. Oxford and Cambridge, Mass.: Basil Blackwell, 1989; Harmondsworth, England: Penguin Books, 1991.

———. Four encounters with Sierpiński's gasket. *Mathematical Intelligencer*. Forthcoming.

Tile and Error

Henry Worm, his wife, Anne-Lida, and baby Wermentrude had survived the trauma of moving into a modern bungahole, all mod cons including bird-alarm. Now feeling much more relaxed, they were beginning to appreciate the advantages of their new home. Anne-Lida in particular was pleased not to have any steep tunnels to climb. But she was *not* pleased at the state of the bathroom.

"Henry! Come *here*, Henry!"

Henry Worm wiggled reluctantly out of his study, which was equipped with a comfortable chair and a shelf of his favorite books, and made his way along the unfamiliar tunnel to the bathroom.

"Yes, my dear?"

"Henry, didn't the builder promise that the bathroom would be fully tiled?"

"Indeed he did, my pet."

"Well, it's not! The wall's been left as plain plaster, and there's a huge box of tiles sitting in the corner!"

"I'll go and phone him right now, dear," said Henry resignedly.

And he did. Only to be told, "Yes, guv. Yes, that's right, guv. Well, had a little bit of a problem. You see, those tiles your missus ordered, the fancy ones with the funny shape, well . . . we can't fit 'em together."

"I beg your pardon?"

"Young Tyler the tiler said he kept getting gaps."

Henry Worm stamped his tail in annoyance. "That's *ridiculous!* I've never heard such a stupid excuse in my life!"

"That's as may be, guv, but it's got us fair flummoxed, I'll tell you."

"Why don't you just fit a few together to make a rectangle, and then fit the rectangles together to cover the wall?"

"Well, yes, guv, Tyler thought of that. It's an old trick of the trade, you know. But the snag was, he couldn't fit 'em into a rectangle. Also, you've got to remember, guv, that making a rectangle ain't the *only* way to tile a wall."

But by now Henry had his heart set on a solution by rectangulation, and his response was prompt. "Nonsense!"

"Look, I'll send him back to have another go," said the builder apologetically. "Ummmm — could fit you in . . . three weeks from next Tuesday."

"Rubbish!" said Henry. "I'll do it myself!" And he slammed the phone down and stomped off. (If you don't know what a worm stomping looks like, imagine a very angry concertina.) He stomped back into the bathroom, offered his wife a few general observations on the competence of the building trade, and opened up the box of tiles.

"Hmmph. Yes, the shape is a bit unusual," he said (Figure 8). "But hardly *difficult*. How to fit them together to fill a rectangle, that's the problem. Pah, it's got to be *easy* . . ."

Two days passed, and Henry Worm became more and more frustrated. Eventually he was forced to take expert advice from his friend Albert Wormstein, who worked in the Patent Office.

"Interesting" said Wormstein, after considerable thought. "I observe that your tile is a form of *polyomino**, that is, a plane figure

---

*The term *pentomino*, for a 5-omino, is a registered trademark of Solomon W. Golomb (No. 1008964, U. S. Patent Office, April 15, 1975).

FIGURE **8**   Henry Worm's bathroom tile. Can you use copies of it to tile a
rectangle?

THE NUMBER OF *n*-OMINOES

n	P(*n*)	Q(*n*)
1	1	1
2	1	2
3	2	6
4	5	19
5	12	63
6	35	216
7	108	760
8	369	2,725
9	1,285	9,910
10	4,655	36,446
11	17,073	135,268
12	63,600	505,861
13	238,591	1,903,890
14	901,971	7,204,874
15	3,426,576	27,394,666
16	13,079,255	104,592,937
17	50,107,911	400,795,860
18	192,622,052	1,540,820,542

Note: $P(p)$ = number of different *n*-ominoes counting
rotated or reflected copies as being the same. $Q(p)$ =
number of different *n*-ominoes *not* counting rotated or
reflected copies as being the same.
It has been proved by David Klarner, John Conway and
Richard Guy that for large *n*, $P(n)$ is approximately
$Q(n)/8$, and that $Q(n)$ is approximately a^n where
$3.72 < a < 4.5$.

formed by joining a set of equal-sized squares edge to edge so that the corners match. If you need n such squares, it's called an *n-omino*." The number of *n*-ominoes increases rapidly with n (see the table on page 21).

"To make the problem precise," Wormstein continued, "let me follow David Klarner, who in 1969 defined the order of any polyomino to be the smallest number of copies that will fit together to fill a rectangle. Assuming this is possible, of course. If not, then the order is not defined."

PROBLEM ❶

Can you think of any polyominoes that do not have an order, that is, will not tile a rectangle? The rectangle can be any shape (as long as it's rectangular!) or size that you like.

Some examples of polyominoes of various orders are shown in Figure 9. Note that to prove a polyomino has order m, say, you must do *two* things:

A. Find a way to tile a rectangle with m copies,
B. Prove that no number of copies smaller than m will also tile a rectangle.

You might think that part A is the harder, but often it's B that causes trouble. For any given size and shape of rectangle and any given polyomino it can, in theory, be decided whether the polyomino can tile the rectangle. However, the only systematic method known is intelligent trial-and-error, and this rapidly becomes impractical for even moderate-sized rectangles. No *efficient* general method is known, and there are many unsolved problems. The field is wide open for the recreational mathematician.

"Now," Albert told Henry, "your tile is a deceptively simple example of a heptomino, that is, a polyomino formed from seven squares. What we really want is to know whether it has an order, and if so, what that order is. But I think the main problem is that it's a bit ambitious to

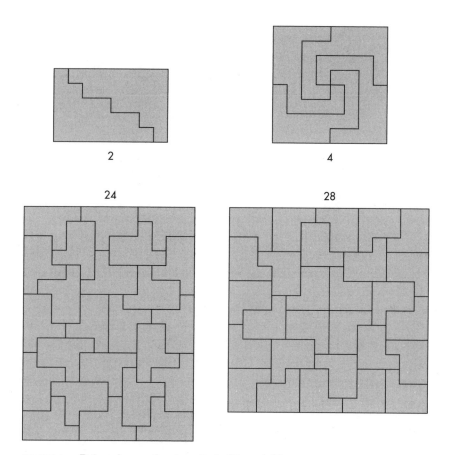

FIGURE **9**   Polyominoes of orders 2, 4, 24, and 28.

start with heptominoes, Henry. Why not start with monominoes, domi-
noes, trominoes, tetrominoes, or pentominoes?"

PROBLEM ❷

The possible n-*ominoes for* n = 1, 2, 3, 4, 5 *are shown in Figure 10.*
Can you work out what their orders are? Here's a hint: the
Y-pentomino may cause you some difficulty, but it can *be done.*

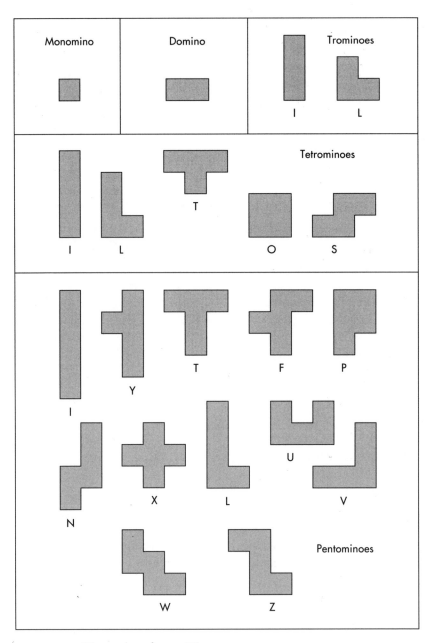

FIGURE 10 All *n*-ominos for *n* ≤ 5.

"Aha! I can spot some general principles even at this early stage," said Albert. "Obviously, a polyomino has order 1 if and only if it is a rectangle."

"If that's the best you can manage, Albert, I think I'll go home and do something more exciting, like watching the cactus growing."

"Merely observing a fact," said Albert huffily. "Even the most trivial facts can be useful. But let us progress to less simple things. A polyomino of order 2, by definition, is obtained by cutting a rectangle into two equal pieces. To achieve this, the cutting line must be symmetric under 180° rotation. This gives an effective characterization of all polyominoes of order 2. It's useful, as well, because most known examples of polyominoes that *have* orders have order 2.

"Order 3 is a more intricate matter. Klarner conjectured in 1969 that polyominoes of order 3 do not exist, adding that 'this idea is intuitively clear, but it seems difficult to formulate a precise proof.' As far as I know, nobody has yet published one: certainly a paper written in 1989 alludes to the problems as unsolved. But Klarner was right, because I, Albert Wormstein, have recently found a proof."

PROBLEM ❸

Can you prove Klarner's Conjecture? Albert's proof is too complicated to give here. Interestingly, it makes extensive use of symmetry arguments. Albert is currently publishing it as a joint paper with me, since mathematics journals tend to be reluctant to accept contributions from worms — an appalling example of how science is infested by speciesism. (How many articles by worms have you seen in mathematics journals?) This is the first conjecture in mathematics to have been proved by a worm.

"What about order 4," asked Henry, getting into the spirit of the inquiry. It was clear that it would take Albert a while to get round to Anne-Lida's bathroom, but Henry was in no hurry. Definitely not. Especially if there *wasn't* an answer, which was starting to look ominously possible.

"It is conjectured that order 4 polyominoes arise in exactly two ways," said Albert (Figure 11). "A proof seems likely to be complicated

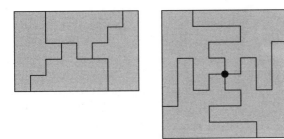

FIGURE 11   Two ways to obtain order 4 polyominoes.

—well, I certainly haven't found one—but not outside the bounds of possibility if you're careful with the bookkeeping."

"And order 5?"

"Nobody knows one. Until recently the only known orders bigger than 4 were 10, 18, 24, 28, 76, and 92."

"Those are very curious numbers," said Henry.

"They certainly are," Albert agreed. "If they were the only ones possible, 92 would be a very unusual number, the largest possible polyomino order."

"Hmmm. Does 92 have any other unusual properties?"

Wormstein pulled a copy of David Wells's *The Penguin Dictionary of Curious and Interesting Numbers* off his shelf. "There's nothing about it in here," he said.

"Perhaps it isn't curious and interesting," said Henry.

"There's an old mathematicians' joke about that," said Albert. "A theorem that every number is interesting. The proof is that, if some number isn't interesting, then there must be a *smallest* uninteresting number. And that, of course, would be a very interesting number indeed! So there's a logical contradiction, implying that the assumption that an uninteresting number exists must be false."

"Ah, yes," said Henry. "But would it also be curious?"

"I don't know," said Wormstein. "That's a very curious question to ask. However, it may be beside the point, because in 1989 Solomon Golomb proved that every multiple of 4 can be an order."

"What about my bathroom tile?" said Henry.

"That's a heptomino, and I haven't even done *hexominoes* yet," replied Wormstein. "Now, there are precisely 35 hexominoes, and it is known that exactly 10 of them tile a rectangle. I'll give you some clues, and you can try to find them. There are two of order 1, five of order 2, and one of order 4." Wormstein quickly took paper and pencil and made a sketch for Henry. "Here's a polyomino [Figure 12*A*] of order 18, and you might like to cut out 18 copies and try to fit them into a 9 × 12 rectangle. Until recently there was exactly one case where the answer was unknown, the Y-hexomino [Figure 12*B*]. But in 1989 Karl Dahlke proved that the Y-hexomino has order 92. I promised to say more about that earlier, and now I have."

PROBLEM ❹

Can you find the eight hexominoes of order 4 or less? Can you tile a rectangle using the polyomino in Figure 12A? You can try to find how to fit 92 Y-hexominoes into a 23 × 24 rectangle if you like, but Dahlke took three days on a microcomputer using the programming language C.

"Yes, Albert. You also promised to say something about Anne-Lida's bathroom tile."

"Of course. That's rather fascinating. You see, again until recently only one heptomino of order greater than 2 was known. The others, with one exception, were the 1 × 7 rectangle, or were given by the constructions I described earlier for order 2, or were known not to have an order. The only high-order heptomino was discovered by James Stuart, of Endwell, New York, and has order 28" (see Figure 9).

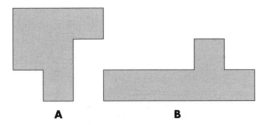

    A   **B**

FIGURE 12   *A.* A hexomino of order 18. *B.* The Y-hexomino, whose order was unknown until 1989.

"Ah! And that one's our bathroom tile?"

"Of course not! No, *your* tile happens to be the only one not accounted for—"

Henry said a very unwormlike word.

"—until *very* recently," Albert finished. "Karl Dahlke has come to your rescue again! Also in 1989 he modified his Y-hexomino program to prove that your heptomino tile has order precisely 78" (Figure 13). (An alert reader pointed out something rather strange about Dahlke's paper: its title claims the order is 76, but the picture that it exhibits as proof uses 78 heptominoes.)

Henry thanked his friend profusely and hastened home to his bunga-hole to tell Anne-Lida. *She'll be enormously impressed!* he thought. When he arrived she was in the kitchen, making a leafmold blancmange.

"Anne-Lida! I've cracked the problem of the tiles!"

"As long as you don't crack the *tiles*, Henry," she sniffed.

"No, really! All you have to do is take 78 of them and form a rectangle!"

"Henry, there are times when you excel yourself. That's wonderful news! We'll start right away. You count out 78 tiles while I clean the

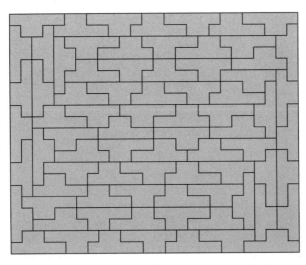

FIGURE 13   The answer to Henry's problem of tiling a rectangle with 78 of his bathroom tiles.

walls down." She swept majestically out of the room and down the hallway to the bathroom. Henry followed.

As Anne-Lida began to scrub the walls vigorously, Henry picked up the carton of tiles. *Maybe the label will say how many are in here,* he thought. *Then I can save effort by just taking a few out.* He turned the box round to see the label. It read:

Combinatorial Ceramic Corporation
Heptomino Tiles
Contents: 77

Henry said another very unwormlike word, but under his breath so that Anne-Lida didn't hear. He turned nervously to address his wife.

"Um. Anne-Lida my sweet, I think we have a little problem . . ."

ANSWERS

1. Perhaps the simplest example of a polyomino that cannot tile a rectangle is the octomino formed by a 3×3 square with a hole. Filling the hole is what causes the trouble.

2. The monomino and domino are rectangles, hence have order 1. So do the I-tromino, I-tetromino, O-tetromino, and I-pentomino. The L-tromino, L-tetromino, L-pentomino, and P-pentomino have order 2. The T-tetromino has order 4, and the Y-pentomino has order 10. (This is the only case where you may have to experiment for a while; use a 5×10 rectangle.)

   The remaining cases — the S-tetromino and the T-, F-, N-, X-, U-, V-, and W-, and Z-pentominoes — do not tile a rectangle. The proof in each case is readily obtained by considering what happens at one corner of a putative rectangle that they tile: the placement of other pieces is forced and prevents anything fitting at the adjacent corner.

   The table on page 30 summarizes these results.

3. Albert's proof will soon be published; see Further Reading at end of this chapter.

4. The ten hexominoes of finite order are shown in Figure 14. The final one is Dahlke's solution for the Y-hexomino.

ORDERS OF n-OMINOES FOR $n \le 5$

	Size Omino	Order
monomino		1
domino		1
trominoes	I	1
	L	2
tetrominoes	I, O	1
	L	2
	T	4
	S	*
pentominoes	I	1
	L, P	2
	Y	10
	F, N, T, U, V, W, X, Z	*

*indicates that the order is not defined

FIGURE **14**   The hexominoes of finite order.

FURTHER READING

Dahlke, Karl A. The Y-hexomino has order 92. *Journal of Combinatorial Theory* Series A 51 (1989): 125–26.

———. A heptomino of order 76. *Journal of Combinatorial Theory* Series A 51 (1989): 127–28.

Gardner, Martin. *Mathematical Magic Show*. New York: Alfred A. Knopf, 1977; Harmondsworth, England: Penguin Books, 1985.

Golomb, Solomon W. *Polyominoes*. New York: Scribner, 1965.

———. Tiling with polyominoes. *Journal of Combinatorial Theory* 1 (1966): 280–96.

———. Polyominoes which tile rectangles. *Journal of Combinatorial Theory* Series A 51 (1989): 117–24.

Grünbaum, Branko, and G. C. Shephard. *Tilings and Patterns*. New York: W. H. Freeman, 1987.

Klarner, David A. Packing a rectangle with congruent n-ominoes. *Journal of Combinatorial Theory* 7 (1969): 107–15.

Stewart, Ian, and Albert Wormstein. Polyominoes of order 3 do not exist. *Journal of Combinatorial Theory* Series A. Forthcoming.

Wells, David. *The Penguin Dictionary of Curious and Interesting Numbers*. Harmondsworth, England: Penguin Books, 1986; New York: Viking Penguin, 1986.

Through the Evolvoscope

The theory of evolution has been a constant source of controversy ever since 1859 when Charles Darwin proposed it in his book *On the Origin of Species by Means of Natural Selection.* That's not surprising: it refers to events long past, debunks humankind's position not only as the pinnacle but also as the *purpose* of creation, and treads on the toes of half the religions on the planet. Many sects have come to terms with Darwinism by considering it as the mechanism employed by the deity to create humanity; but others — such as the "creation scientists" in the United States — have taken political action to promote their own theories in place of Darwin's. In the 1980s Arkansas and Louisiana passed legislation requiring equal treatment of evolution and creation science in schools, even though the latter had not been subjected to the same lengthy scientific scrutiny as the former. After considerable protest, the legislation was overruled by the Supreme Court as a violation of the constitutional separation of church and state. It is important to recognize that Darwinian evolution is just a scientific theory, not a definite fact; but it is also important to recognize that scientific theories are generally tested far more stringently than the alleged facts of religions or philosophies. Research into evolution continues to be active, and many problems remain unresolved, as we shall shortly see.

Darwin's basic idea — that a combination of random mutation and competition between reproducing individuals inevitably leads to an evolutionary process and the formation of ever more complex species — has an elegant simplicity. So simple is it that some people consider it tautologous (which in a sense it is) and therefore lacking explanatory power (which it does not). But science does not make progress by accepting ideas just because they are elegant; it does so by subjecting them to the most rigorous tests possible and comparing their predictions with actual events.

With evolution this isn't so easy. The "predictions" refer to the distant past, and testing them is largely done through the fossil record, which is about the worst kind of database that any scientist would wish to work with: unreliable, hard to interpret, full of gaps, and liable to revision at any moment. This, of course, is what makes it all such jolly fun and fuels the fires of controversy.

One of the controversies currently ranging — and only one, for there are others, all mixed up together — is between what we might call the punctualists and the gradualists. The gradualists, who (with some simplification) may be represented by Richard Dawkins, hold that evolution is a continuous process that occurs as a long series of very tiny steps. The punctualists, of whom the best known is Stephen Jay Gould, claim that the evolution of a new species occurs only in rapid bursts. We've all seen those elegant evolutionary trees in the textbooks (Figure 15) whose bifurcating branches sweep majestically skyward to culminate in *Homo sapiens*, perched triumphantly at the top like a star on a Christmas tree. These reconstructed trees are very tidy, but the fossil record on which they are based is much more jerky and irregular. The gradualists explain such apparent discontinuities as gaps in the fossil record, perhaps to be filled by future discoveries of new fossils, or destined to remain unfilled because of the erratic manner in which rocks are deposited. It must be said that this is a not unreasonable view. The punctualists, on the other hand, believe that the discontinuities in the

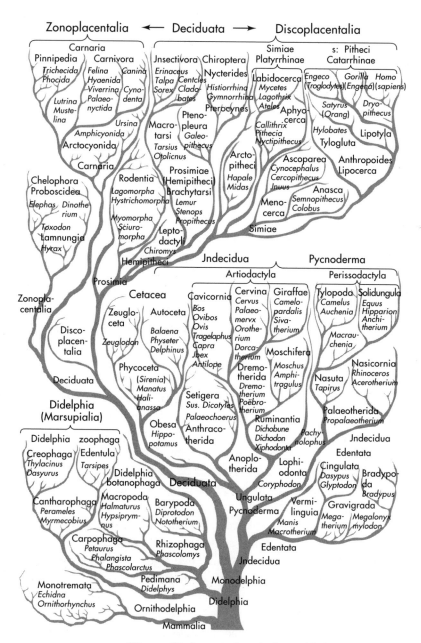

FIGURE 15 The textbook tree of evolution, with humanity at its peak, this one from *Generelle Morphologie der Organismen* by E. Haeckel, 1866.

fossil record are genuine and are in fact the sole source of new species. Who is right: the punctualists, the gradualists—or neither?

One evening recently I was relaxing on the couch, cat on lap, and musing idly about the problem of gradual and sudden changes in evolution, when the doorbell rang. It was a messenger with a parcel. The parcel was wrapped in plain brown paper, with a blue label reading "EcoMat Instrument Company Inc." and a second red label reading "FRAGILE!" This was puzzling, since I had never heard of the EcoMat Instrument Company and certainly had not ordered anything from it. However, it is fruitless to puzzle over the outsides of sealed containers, so I removed the wrapping and found a cardboard box labeled "Evolvoscope: Mark I prototype."

My curiosity now thoroughly aroused, I opened the box to discover a curious piece of headgear, not unlike a cross between a personal stereo and sunglasses. The control box was covered in dials, buttons, knobs, levers, and flashing light-emitting diodes. Yes, it looked *just* like a personal stereo.

A thick manual, ring-bound in a plastic cover, lay beneath it.

I picked up the apparatus and settled it on my head. There was a strange feeling of discontinuity and the room swam for an instant; then it clicked back into focus, and everything was different. Instead of the room, there was a flat plane. Actually it wasn't really a plane—the harder I concentrated the more dimensions I could see, hundreds, thousands, millions . . . But I kept thinking of it as a plane. It was populated by the most peculiar creatures I have ever seen, crowded densely together. I looked towards where my lap should be and there, indeed, I saw my cat, a sleepy bundle of brown fur. But next to it was another cat, and another, and another—millions of cats, packed tighter than any sardines . . . and all subtly different from one another. Some were tabby, some black-and-white; some fat, some thin. As I cast my eyes further to one side I saw cats with two tails, five legs, three eyes; a green cat with roller-skates instead of paws, a transparent cat; a spherical cat that rolled along the ground in pursuit of an equally spherical mouse . . .

How did they fit into such a small space? Puzzled, I flipped open the manual. It was about as well written as manuals ever are, that is, impenetrable—and with difficulty I found an entry that appeared relevant:

Form space. Default mode for the Evolvoscope is to present a static view of real-time form space. Each point in this multidimensional continuum corresponds to the form of a potential living creature. The evolvoscope permits direct viewing not only of the space but also of the creatures associated with each of its points. To select for survival capability, depress the button labeled FITNESS FUNCTION on the control unit.

I found the button and pushed. The plane began to swell into a rounded, hillocky landscape (Figure 16). My cat floated before my nose, perched upon a small bulge. In the distance were dogs, a cow, a giraffe, and an earthworm, each perched upon its own hillock. There were other, stranger creatures: one I named a telephant, because it resembled an elephant with a telephone-shaped trunk. There was a goatilus (a goat with a spiral shell on its back) and a cricket bat (a cross between a

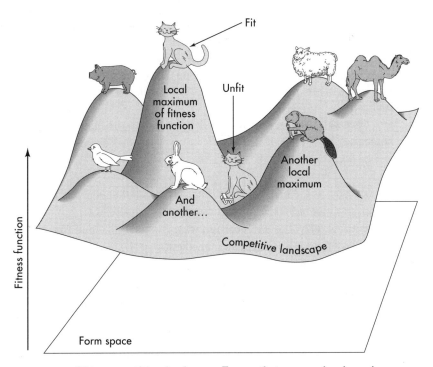

FIGURE 16   The competitive landscape. Forms that occupy local maxima are fit to compete and survive; forms that occupy valleys are unfit and die out.

cricket and a bat, naturally). There was a chimpanzebra and a swan-garoo, a frogodile and a flipperpotamus. I looked for the roller-skated cat at first without success, until I peered into a deep valley off to one side, where I found it lurking, surrounded by others of its kind. Indeed, I soon perceived that the most familiar creatures all seemed to occupy the tops of hillocks, and the more bizarre animals inhabited the valley floors. As my eye scanned the slope from floor to peak, the creatures became gradually more familiar: the hilltops were populated exclusively with animals that I could recognize as normal inhabitants of our planet. Some hillocks, however, seemed unoccupied.

I thumbed through the manual, and light began to dawn.

> *Fitness function.* Obtained by depressing the corresponding button on the control unit. This displays a measure of fitness of any given creature in form space, compressed schematically into a one-dimensional quantity, represented by *height*. Thus fitness is represented as a graph of this quantity against points in form space. This has the effect of displacing form space in the vertical direction by an amount proportional to the fitness function: mathematically this corresponds to displaying the graph of the fitness function. The default plane for form space becomes curved, and now represents the *competitive landscape*. Species that are more fit to survive occupy higher levels in this landscape.
>
> *Optional zoom facility.* By twisting the dial on the bridge of the nosepiece, local scale may be varied to render microscopic forms visible.

So *that* was why some hillocks were empty! I zoomed in on a few of them and found an amoeba, a hydra, and a salmonella bacterium.

I now knew what the evolvoscope was. It was a machine that made visible the various mathematical ingredients of Darwinian evolution. Although Darwin couched his theories verbally, they can be translated into mathematical concepts, making their consequences easier to analyze unambiguously. Of course, any given mathematical model of Darwinian evolution is only an approximation to the real thing—assuming there *is* a real thing, which of course is as controversial an assumption as everything else in this subject—but precise mathematical formulations help us to distinguish between variant interpretations. They also can reveal inconsistencies in logical analysis by acting as counterexamples to certain types of verbal argument. The trouble with words is that they can subtly change their meanings as a logical argument progresses, and the conclusions reached may thus be invalidated. Mathematics is more precise, and it lets us pin down the exact assumptions being made.

In the particular model of evolution that was built into the evolvoscope, animal forms were represented by a point in form space. That space contained not only all viable forms, but also all *imaginable* forms. There was also a measure of fitness to survive, the fitness function, and the forms that could compete successfully were those whose fitness levels were as large as possible. Species corresponded to local maxima of the fitness function — hillocks in the competitive landscape, or environmental niches (Figure 17).

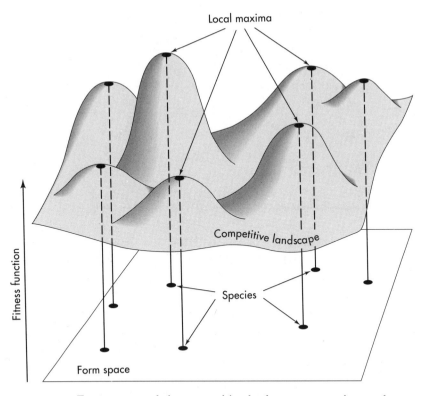

FIGURE 17   Form space and the competitive landscape are continuous, but species correspond to local maxima of the landscape and are thus discrete.

This helps to explain why, given a continuous range of possible forms, species tend to be clearly separated from each other. Mice, for instance, are all fairly small: you don't get a broad and continuous range of sizes of mouselike creatures. Form space is continuous, but each local maximum is surrounded by its own hillock, and the hillocks are separate from each other; both the niches and the species that occupy them are discrete.

Another phenomenon that fits in well is convergent evolution, the occurrence of forms that are similar but that have quite different genes. For example, there is an astonishingly close parallel between the marsupials of Australia and the placental mammals of the other continents:

Marsupial	Placental Mammal
Tasmanian wolf	Wolf
Flying phalanger	Flying squirrel
Marsupial mouse	Mouse
Marsupial mole	Mole
Banded anteater	Anteater
Wombat	Groundhog

PROBLEM ❶

Think of two further examples of convergent evolution.

At that stage I stopped fiddling with the evolvoscope — I had no idea what features (or creatures) might be built in, and had no wish to seed the planet with mutated man-eating cucumbers, or whatever. Before going any further, I read the entire manual from start to finish.

Then, satisfied that the device produced purely a simulation, and that nothing I did would affect the actual terrestrial environment, I started to experiment. The next manual entry got to the heart of Darwin's argument.

For a more accurate representation of the true situation, turn the knob marked RANDOM MUTATION. WARNING! If the level of random muta-

> tion is turned up to high, chaos may result. Experiment by turning the
> knob *gradually*.

Gingerly I turned the knob; my cat turned ginger. Then tabby, then siamese, then manx. I turned the knob a little further, and the cat became fuzzy. It was *vibrating*.

I depressed the SLOW MOTION lever to see the action more clearly. Every few seconds, my cat would change its form slightly. As it changed, it also moved over the competitive landscape. There were all manner of cats clustered on the hillock, changing form and moving around in a bewildering dance. An occasional big change moved the cat further than normal, but usually it would drift back to the top of its hillock. For example, temporarily it grew ears like an elephant's; but I noticed that it then slid some way down the side of the hillock. Presumably elephant ears were an unfavorable mutation, rendering the cat less fit to compete. The sheer effort of dragging its ears around would slow it down considerably, and mice would hear the ears flapping as the cat lumbered towards them.

Mutations that produced cats lower down the hillock didn't last. The cats drifted upwards again, changing back towards the more normal feline form as natural selection weeded out the less orthodox shapes.

I turned the MUTATION knob further. The hillock seethed with activity, vibrating cats surging around like waves in a stormy sea. Suddenly, with an audible PLOP!, a blob full of cats slid down the hillock, hesitated in the valley and shot up a nearby hill that until now had been empty (Figure 18).

I slowed the action down again and inspected the newcomers.

Fascinating.

The world's first flying cat.

PROBLEM ❷

On the face of it, a flying cat would appear fitter to survive than one without wings. Explain why the nonexistence of flying cats in the real world is not inconsistent with Darwinian evolution.

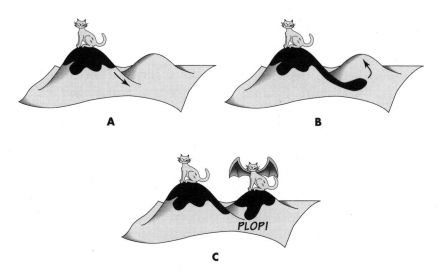

FIGURE 18   Effect of random mutation on creatures occupying an environmental niche. *A.* An ensemble of possible forms (dark blob) confined within the niche. *B.* Random changes cause the ensemble to put out a feeler towards a nearby unoccupied niche. *C.* Part of the ensemble splits off to form a new species.

I realized what I had witnessed: *the evolution of a new species.*

Random mutations, at a very high rate, compressed into a few minutes, had produced a new form of living creature, one that — unlike most mutations — was fit to survive. The ensemble of possible cat-forms had slopped over into a new environmental niche, a new hillock in the competitive landscape, and occupied it. Random mutations, acting as "noise," had pushed a section of the cat species out of one local maximum and into another!

Did that explain the sudden changes in the fossil record?

Not really.

First, mutations of the type I had just witnessed would be very unusual; this is in fact one standard objection to Darwin's theories. The whole series of changes of form that would be needed to create a truly new species in one go would be very rare. I had either been lucky or I had turned up the mutation rate too high.

Moreover, when species change suddenly in the fossil record, what often seems to happen is that you get several new species occurring at once. Instead of bifurcation (splitting into two) of the branches of the evolutionary tree, you see multifurcation (multiple branches occurring simultaneously). Darwin realized this: in fact, there is only one diagram in *The Origin of Species*, and it illustrates this precise point. If the evolution of a new species occurs by representative creatures slopping over randomly from one niche to another, then it should produce only one new species on each occasion.

Finally, with this mechanism, the original species remains: no species ever dies out. Not very plausible. Of course, the entire blob of cats might slop over into the winged cat niche, but that seems even less likely than that part of it might do so. Not very plausible.

What I wanted to find in the mathematical model was a more robust method for the evolution of new species. Having read the manual, I was soon on a promising line of inquiry.

> *Changing environment.* In the real world, the level of fitness of a creature depends on the environment. For example, a creature that has evolved a long neck to eat leaves from tall trees will lose its competitive advantage and thus become less fit if drought reduces the height of trees. The CHANGING ENVIRONMENT lever permits the user to experiment with different kinds of change, either in time or in space.
>
> It is suggested that on the initial attempt at changing the environment, the random mutation rate should be set to zero or to a very small value.

Right, I thought. *Let's try time-variation.* (Spatial variation is also interesting, but I'll let you experiment with it.) I pushed the lever a little.

The landscape began to undulate like a slow-motion film of a stormy sea. The groups of viable creatures clung to their environmental niches, staying perched on the top of their particular hillock as it bulged, shrunk, rose, fell, or moved bodily sideways. The species was able to respond rapidly enough to change that it could continue to maximize its competitiveness, its fitness to survive. I realized that I was observing the inherent *stability* of particular species: in a slowly changing environment, the form of a species changes slowly, if at all. Evidence, it seemed to me, for the gradualist theory of evolution . . .

But what was *that*? Out of the corner of my eye, I had seen sudden, violent action. I continued watching the rippling landscape, and then I saw it happen. An entire species suddenly whizzed up a nearby hillock, changing form in an instant as it occupied a new niche.

Why had it moved from its old niche? Come to that, where *was* the old niche?

On television, I thought, it would now be time for the slow-motion replay. Then I'd be able to see what really—hang on, *another* button?

It was marked SLO-MO REPLAY. Fair enough. I pushed it.

Now the sequence of events was obvious (Figure 19). The rapid motion occurred when an environmental niche *disappeared*. First there was just one niche: a low, occupied hillock. Then, nearby, an unoccupied one began to grow from a previously featureless part of the landscape. The changing environment had created the potential for a new species. But the new niche was unoccupied, because no creatures had managed to make the transition needed to occupy it. The new niche grew, and soon it was higher than the old one. It moved towards the old one. Between them was a narrowing pass, a saddle in the surface. The saddle rose and met the old niche—and the two canceled each other out, like a particle hitting an antiparticle. Only instead of releasing energy, the collision released the creatures from the trap of the old niche. Now free to improve their fitness value, they shot up the new hillock at high

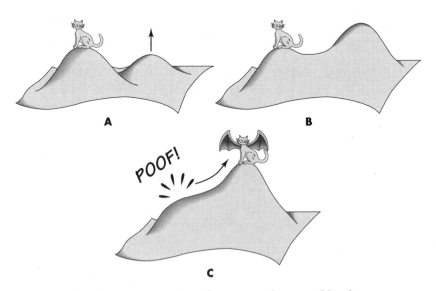

FIGURE 19   Catastrophic evolution of a new species caused by the disappearance of a niche.

speed, jostled around a bit, and occupied the top. But now they looked different: they had evolved to fit the new niche.

They'd done it quickly, too! So this time it looked as though the punctualists were right.

Maybe they were *both* right! It seemed to me that there were two distinct phenomena. Both had the same general cause — environmental variation — but happened in distinct circumstances. If the change is slow and continuous, the environmental niche changes shape or moves, but remains the same niche; in fast, effectively discontinuous change, an environmental niche collides with another and disappears.

That was more like it! The interesting feature is that both gradual change and punctuated equilibrium emerge as natural consequences of the *same* mathematical model of Darwinian evolution. Moreover, even if the mathematical model does not correspond to actual evolution, it still demonstrates that it is *possible* for both types of change to coexist, without having special and distinct causes, and that there is no logical necessity to consider them as conflicting theories. In other words, perhaps the gradualists and the punctualists should put their heads together instead of banging them against each other.

The book had a footnote:

> The mathematical theory of singularities, otherwise known as catastrophe theory, invented by René Thom, Vladimir Arnold, John Mather, and others, shows that the two types of behavior observed above are the typical things — and the *only* typical things — that can happen in this kind of mathematical model. The difference is that slow change happens most of the time, whereas sudden change (or catastrophe) occurs only at isolated instants. However, this is precisely what happens in the fossil record. Christopher Zeeman applied catastrophe theory to evolution, following the work of Maurice Dodson and others.
>
> Oddly, the earliest proponents of sudden changes in evolution, back in Darwin's day, were called catastrophists.

At this point I saw that not all difficulties were yet solved. The type of sudden jump occurring in the model — and according to catastrophe theory the only possible type of jump, save in highly unlikely exceptional cases — involves the *disappearance* of a single old species and the creation of a single new one. Which didn't explain Darwin's observation of the creation of several species at the same time, all stemming from the same precursor.

After a while it occurred to me that I'd turned off the RANDOM MUTATION feature. Maybe that made a difference, though I couldn't see how.

I ran the experiment again but with RANDOM MUTATION turned on. Now, instead of a single form perched on the top, each niche had a vibrating blob of related but dissimilar forms, occasionally slopping some way downhill but generally confined near the hilltop by selection of the fittest. The first try produced nothing very different; when the niche disappeared the whole blob shot off up the new hill and ended up in the same place as before.

But I noticed that as the blob moved up the hill it tended to spread out sideways. There's a good reason why this happens. Think of a particular form for a creature, surrounded by a circular blob of random variations on that form. Now place that blob on a slope. The forms on either side are at the same level of fitness, lying on the same contour of the competitive landscape. So they are equally fit, and hence survive at the same rate. The forms lower down the slope are less fit, and hence less able to survive. Forms higher up the slope, on the other hand, are even fitter to survive. So the circular blob is compressed from behind but can spread out sideways: it tends to resemble an ellipse, flattened along the lower edge.

Suppose a blob of possible forms moves up a slope that is near not just one hillock, but several? Then as it spreads sideways it can break up into pieces that occupy *all* of the nearby hillocks. And that leads to multiple speciation—just as Darwin observed. The corresponding evolutionary tree looks like Figure 20 *A*; but because rock-producing sediments are laid down very slowly the timescale is effectively compressed, and what we see in the fossil record is more like Figure 20 *B*—which is indeed what we *do* see. It all fits!

So when we put both of the essential ingredients of Darwin's theory together—random mutation *plus* survival of the fittest—then the simplest, most natural mathematical model has all of the following features:

- Most of the time species change slowly or not at all.
- During slow changes, no new species are created.
- On isolated occasions species can change much more rapidly.
- When sudden, catastrophic changes occur, multiple species often evolve simultaneously.

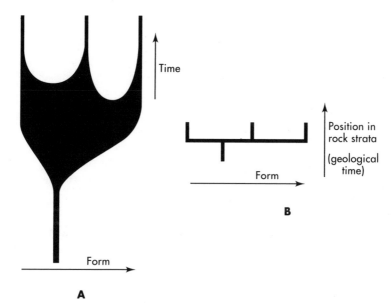

Time

Position in
rock strata

(geological
time)

Form

B

Form

A

FIGURE **20**   *A.* Evolutionary tree corresponding to multifurcation, as in
Figure 15. *B.* In the fossil record the same tree appears highly compressed,
yielding apparent discontinuities. Compare with Figure 16.

Notice that the catastrophes are *not* just speeded-up versions of the slow
changes; qualitatively, they are entirely different. Mathematically they
involve the disappearance of a niche, whereas the slow ones involve only
the continuous modification of a niche. Moreover, even a speeded-up
"slow" change cannot produce a new species: only a catastrophe can do
that. All this is sufficiently close to what we see in the fossil record that
it really is doubtful whether gradualism and punctualism are truly conflicting theories. Maybe they are just two distinct aspects of the *same*
theory! Not gradualism, not punctualism — but both.

Of course, this is just the simplest model. There are many ways,
other than catastrophe theory, to translate Darwin's hypotheses into
mathematics. But more complicated, more realistic models should just fit
the fossil record better. They can still combine slow and fast change in
one consistent theory. Indeed, the relative success of the simple idea just

outlined strongly suggests that they *should,* that this combination is a natural consequence of general mathematical principles, rather than being a puzzle.

The manual ended with a caution:

> WARNING! The evolvoscope Mark I is a prototype. More advanced models will provide sophisticated multivariate measures of degree of fitness rather than one-dimensional numerical measure. The competition between individuals will affect the environment, to model the possibility that changes in the competitors may change the rules of the competition.

That was an excellent point. The evolution of a winged cat would have a dramatic effect on the competitive landscape for birds!

> Proper attention will be given to the distinction between the *genotype* (the creature's genetic material) and the *phenotype* (its form). Natural selection acts on the phenotype, but random mutations act on the genotype. Therefore the model must incorporate the interaction between these factors via the reproductive process . . .

My head was swimming with vibrating blobs of flipperpotami wiggling around on rippling landscapes. I tried to pull the evolvoscope from my head—but there was nothing there.

It had all been a dream.

But dreams and mathematics both exercise the faculty of imagination. And I realized that the evolvoscope—or rather, something like a highly sophisticated version of it, maybe Mark XXVI—*does* exist, after all. It's been around for the last four billion years or so. But it's an invisible mechanism, which operates over periods of millions of years, and we see only its effects.

You want to see a *real* evolvoscope?

You're living in it.

ANSWERS

1. The fishlike shape of the dolphin; the wings of bats and birds.
2. Here are three answers. *A.* A flying cat might *not* be fitter than one without wings. Although wings would help it to catch birds and

escape from dogs, they would hamper it when the cat climbs trees and add extra weight when it chases mice. *B.* The necessary mutations to create a flying cat have not yet happened; the chance element of evolution has been at work. *C.* There are constraints, from physics and chemistry, that affect what *can* evolve. An attribute does not appear in a species simply because it might prove useful; there has to be an effective developmental route.

FURTHER READING

Arnold, V. I. *Catastrophe Theory*, 2d ed. New York: Springer-Verlag, 1986.

Darwin, Charles. *The Origin of Species*. New York: Viking Penguin, 1982; Harmondsworth, England: Penguin Books, 1985.

Dawkins, Richard. *The Selfish Gene*, 2d ed. New York: Oxford University Press, 1990.

————. *The Blind Watchmaker*. Harlow, England: Longman, 1986; New York: W. W. Norton, 1987.

Dodson, M. M. Quantum evolution and the fold catastrophe. *Evolutionary Theory* 1 (1975):107–18.

Gould, Stephen Jay. *Ontogeny and Phylogeny*. Cambridge: Harvard University Press, 1977.

————. *Wonderful Life: The Burgess Shales and the Nature of History*. New York: W. W. Norton, 1989; London: Hutchinson, 1990.

Maynard-Smith, John. *Evolution and the Theory of Games*. Cambridge and New York: Cambridge University Press, 1982.

Stewart, Ian. *Oh! Catastrophe*. Paris: Belin, 1982.

————. *The Problem of Mathematics*. Oxford and New York: Oxford University Press, 1987.

Zeeman, E. C. Decision making and evolution. In *Theory and Explanation in Archaeology*, ed. C. Renfrew, M. J. Rowlands, and B. A. Seagraves-Whallon. New York: Academic Press, 1982.

4

Cogwheels

of the Mind

It was spring in the historic English university town of Cambridge. On the backs, the grassy area beside the river, daffodils splashed great strokes of yellow onto a green canvas. I walked across the wooden bridge that, according to legend, was designed by Isaac Newton and assembled without nails. On one occasion it was taken down, and when the time came to put it back up, nobody could work out how to fit it together again except by knocking in the odd nail or two. Newton's ideas are often hard to reconstruct.

The city is steeped in mathematical tradition. Newton was at Cambridge University, at Trinity College. At nearby St. John's College, in 1845, John Couch Adams performed calculations that led him to predict the existence — and the position — of the then unknown planet Neptune. (The French astronomer U. J. J. Leverrier independently came to similar conclusions. The French were quicker to point their telescopes where the theoretician told them to, and Leverrier rightly gets the credit for the discovery. A similar sense of urgency on the part of the British Astronomer Royal, Sir George Airy, would have given Adams priority.) Cambridge was where the self-taught Indian genius, Srinivasa Ramanujan, first met European mathematicians. Alan Turing, one of the fathers of the computer, was at one time a fellow of King's College. John Horton Conway, famous among mathematical visionaries as the inventor of the game of "Life," was at Cambridge until recently, when he brain-drained to Princeton. Mathematicians remember his invention of the Conway simple group $\cdot O$ ("dot-oh") composed of $2^{22}.3^9.5^4.7^2.11.13.23$ rotations in 24-dimensional space, and his own proudest achievement is being able to turn his tongue into the vertical position.

And two past presidents of Gonville and Caius College have been mathematicians, John Venn and Sir Ronald Fisher. Sir Ronald, the famous statistician, was president from 1956 to 1959. John Venn, the logician who invented Venn diagrams, was president between 1903 and 1923. I was in Cambridge to take a look at a stained glass window, newly installed in the hall of Caius College, to commemorate the two great men. More accurately, these were two windows, one above the other: three overlapping discs of yellow, blue, and purple for Venn, and a 7×7 multicolored checkerboard for Fisher.

Well, that was the ostensible reason. In actual fact I was there to visit an old friend, Lucretia Borges, who like Venn is a mathematical logician and who similarly inhabits the Caius cloisters.

We were sitting at the high table among the fellows, feasting on roast wildfowl in orange sauce and consuming liberal quantities of claret. Fellows are permitted up to two guests, and Lucretia's husband, Jules, had joined us. Jules is a very pleasant but verbose philosopher. He was in full flow, giving a lengthy dissertation on the use of visual imagery in the works of Nicholas of Cusa:

". . . and this would interest you, Ian, because you're a mathematician, aren't you, you see, Nicholas was very, and I mean very, fond of mathematical symbolism. I mean, he used the circle as an image for, well . . . God. You see, if a circle gets bigger than its curvature, the amount it bends, you know, gets less . . . This illustrates how contradictory aspects of the divinity coincide in the infinite; an infinite circle is a straight line, you see, but man must be forever satisfied with an approximation to the infinitude of God."

"That's fascinating, Jules," I said. "Yes, please, I'll have the carameled pineapple in rum sauce . . ."

"Every age," he went on, "seems to have its own store of, well, visual icons, kind of Ur-images, fundamental forms, you know, Plato's cave and all that stuff . . . universally valid symbols."

"Like the Mandelbrot set," said Lucretia. "A fractal. The late twentieth century's icon of chaos."

"Oh, absolutely! Right! And those two diagrams in the stained glass window up there, you see, those are icons too. Their symbolic roots penetrate deeply into the common subconscious of, like, humanity, so that — "

"Jules," I said. "If the symbols are *universally* valid, then Nicholas of Cusa's use of the circle as a symbol of God must mean that Venn's diagram must be a symbol for *three* gods."

"The Holy Trinity," he said with an airy wave of the hand. "Three *linked* circles, you'll note. It would make a better symbol for *Trinity* College, actually — "

"Well, you know, Jules, that's not *quite* what Venn invented them for," said Lucretia. "I don't think he had the Trinity in mind."

"Well, then, what *do* his three circles represent?"

"Sets," Lucretia said firmly.

"Let me explain," I said. "Lucretia knows too much about mathematical logic and she'll make it sound too complicated, be on to prenex normal form in no time, I'll bet. Jules, what do the following items have in common: black cat, crow, fur hat, umbrella, polar bear, chameleon, blonde wig, and beachball?"

"Let me see . . . you can buy them all in Harrod's! No, I withdraw that, I don't think Harrod's sells crows, not much call for them nowadays. I know! You can't use any of them to travel to the Canary Islands!"

"Close," I said. "I suppose you might try hitching a ride on a passing crow. But wrong. Try again."

"I give up."

"What they have in common is: they have absolutely nothing in common," I said. My lucid explanation was missing its target, I could tell. "Let me clarify," I said carefully. "The eight items I have listed — and the number eight is very significant here, you'll soon see — "

"Oh, splendid," cried Jules. "I *love* numerological significance! I mean, in Plato's *Republic* the number 729 symbolizes the difference between the king and the tyrant—"

"The eight items embody all possible combinations of three attributes: blackness, life, and hairiness," I told him.

"Oh, my, I should have thought of that," he said.

"The cat, for instance, is black, alive, and hairy. The crow is black, alive, but *not* hairy. The blonde wig is not black, not alive, but *is* hairy. And the beachball is nonblack, nonalive, and nonhairy. In fact," I went on, "for each of the three attributes *blackness*, *life*, and *hairiness*, I can classify the eight items into four that have the attribute, and four that do not. Thus cat, fur hat, polar bear, and blonde wig are all hairy, but the others are not. I can enclose objects with like attributes in a circle [Figure 21]. Or—as in John Venn's masterstroke of genius—I can simultaneously draw three circles, one for each attribute, and thereby display all possible combinations of the three [Figure 22]."

"Magnificent!" said Lucretia, looking at my drawings.

"I don't quite see what this has to do with God," said Jules.

"It doesn't."

"Unless you're thinking of categories of God's creation," he continued, undaunted. "I mean, like, 'black, alive, and hairy, He created them, and He saw that it—I mean, they—was—I mean were—good.'"

"The circles are simply visual devices," I pointed out. "To group together objects with similar attributes."

"Are you *sure*?"

"Positive."

"Ah. Logical-positivism . . . But Venn shouldn't have used circles, you know. Wrong icon. Too confusing. Makes everyone think of God. Misleading, that."

"If you say so, dear," said Lucretia, a trifle wearily. "Ian, draw them a bit wobbly from now on, eh?"

"They're wobbly already, Lucretia. It's the effect of the claret."

"What's so clever about three overlapping circles, anyway?" asked Jules.

"Well, the idea is that different combinations of the three attributes are represented by different areas within the diagram," said Lucretia. "For instance, think of De Morgan's Law." She scribbled rapidly on a napkin:

FIGURE 21   Grouping objects according to the attributes *black, alive, hairy.*

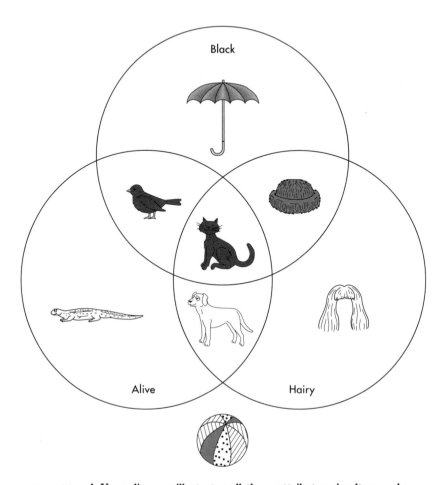

FIGURE **22**   A Venn diagram illustrates all three attributes simultaneously.

$$\neg(p \wedge q) = (\neg p) \vee (\neg q)$$

"I'd rather not, if you don't mind," said Jules.

"Here p and q are logical propositions," said Lucretia helpfully. "The symbol ¬ represents 'not', so ¬p means 'not-p.'"

"To p or not to p," I muttered under my breath. "That is the question."

"Ian, did you say something?"

"Not really, Lucretia, my dear."

"Good. And '$\wedge$' means 'and' and '$\vee$' means 'or.' De Morgan's Law says that 'not-(p-and-q) is the same as (not-p)-or-(not-q).'"

"Say that again, Lucretia, will you?"

"She's making it too complicated, just as I said," I said. "Jules, suppose p is 'black' and q is 'hairy.' Then De Morgan's Law states that anything that is *not* black-and-hairy must *either* be not-black *or* not-hairy."

He digested this remark in silence.

"That's obvious!" said Jules finally. "Isn't it?" he added uncertainly.

"Well," I said, "you might imagine that anything that is *not* black-and-hairy must *both* be not-black *and* not-hairy.

"That is," said Lucretia,

$$\neg(p \wedge q) = (\neg p) \wedge (\neg q)$$

"But that's silly," said Jules. "I mean, a blonde wig is not black-and-hairy, but it's not not-black *and* not-hairy, is it? You know, I mean a wig is *hairy*, not not-hairy, isn't it? And a blonde one isn't black, so it can't be black-and-hairy . . . I'm not putting this very clearly, am I?"

"You're doing fine," I said. "And what you've just said can be drawn on the Venn diagram [Figure 23*A* and *B*]. The blonde wig *does* lie in the region that corresponds to *not black-and-hairy*, but it doesn't lie in the region that corresponds to *not-black and not-hairy*. So those regions are different, which means that the corresponding combinations of attributes are not logically equivalent to each other."

"But on the other hand," said Lucretia, "the region for *not-black or not-hairy* is exactly the same as that for *not black-and-hairy* [Figure 23*C*]. Which shows how you can use Venn diagrams to prove logical equivalences, in this case De Morgan's Law."

Jules drew a deep breath. "You mean it's not about the Trinity at all?"

"No, dear. They're *wobbly* circles, remember?"

"Oh, right. Super."

"Now suppose we were interested not in three, but in four attributes," Lucretia continued. "Blackness, life, hairiness, and —er—harmoniousness."

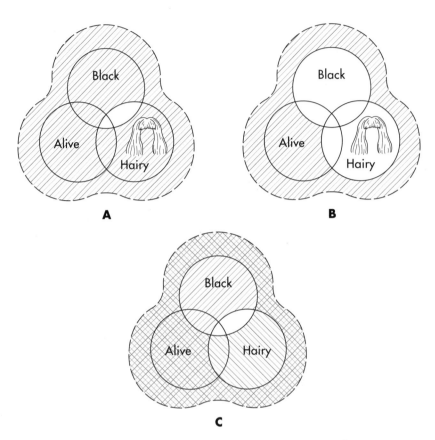

FIGURE 23   A. The region corresponding to *not black-and-hairy*. B. The region corresponding to *not-black and not-hairy*, which differs from A: for example, it does not contain the wig. C. The region for *not-black* (gray) *or not-hairy*, which together cover the same regions as A. This gives a visual proof of De Morgan's Law.

"Eh?" said Jules and I together.

"An umbrella," said Jules, "is black, non-alive, non-hairy, and non-harmonious. A clarinet, on the other hand, is black, non-alive, non-hairy, and harmonious."

"What's black, alive, hairy, and harmonious?" asked Jules.

It's wonderful how the old elephant jokes keep going. After we'd finished falling about, Lucretia gathered the threads of her thoughts.

"What kind of diagram do you need to represent all possible combinations of *four* attributes?"

"Four circles," said Jules immediately.

"No, that doesn't work," said Lucretia. "There are sixteen possible combinations of attributes, but only fourteen regions. There's no region for anything that's black, non-alive, hairy, and nonharmonious, for instance" (Figure 24*A*).

"Like the fur hat," said Jules.

"John Venn knew the right answer," said Lucretia. "He used four *ellipses*" (Figure 24*B*).

"I *told* you it was silly to use circles," said Jules. "Stands to reason, too easily confused with God. I wonder what an ellipse signifies in Nicolas of Cusa's—"

"You're not the only critic," said Lucretia. "In 1896 Lewis Carroll, who had his own version of Venn diagrams, said:

My Method of Diagrams *resembles* Mr. Venn's, in having separate Compartments assigned to various Classes, and in marking these Com-

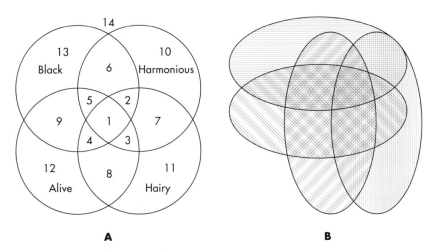

A   **B**

FIGURE 24   *A.* Four circles provide only 14 regions, but there are 16 possible combinations of four attributes. Here the combinations *black, non-alive, hairy, and non-harmonious* and *non-black, alive, non-hairy, and harmonious* are missing. Four circles therefore do not provide an adequate Venn diagram for four attributes. *B.* But, as Venn pointed out, four ellipses do.

partments as *occupied* or *empty;* but it *differs* from his Method, in assigning a *closed* area to the *Universe of Discourse,* so that the Class which, under Mr. Venn's liberal sway, has been ranging at will through Infinite Space, is suddenly dismayed to find itself "cabin'd, cribb'd, confined" in a limited Cell like any other Class!

Carroll suggested a different range of diagrams, using rectangles."

PROBLEM ❶

Find arrangements of three and four rectangles that make effective Venn diagrams.

"Do diagrams exist for every possible number of sets?" asked Jules.

"Spoken like a true mathematician, my dear!" said Lucretia. "Venn himself got stuck at five sets. That is, he never found a really satisfactory answer. In 1880 he did come up with a diagram for five sets, but some of the regions fell into several disconnected pieces. In 1881 he found a way to incorporate a fifth set in the form of an annulus, a circle with a hole in the middle, but wasn't entirely happy with that either. He said that a solution should 'employ only symmetrical figures, such as should not only be an aid to reasoning, but should also be to some extent elegant in themselves.' On the other hand, he also said that 'there is no theoretic difficulty in carrying out the scheme indefinitely.'"

"I can see that," I said. "You just make each new set poke out fingers into all of the previous regions. But then the sets will soon get rather irregular and wiggly."

"Venn knew that, too," said Lucretia. "He said that 'there is a tendency for the resultant outlines thus successively drawn to assume a comblike shape after the first four or five.'"

"Right!"

"Now all this leads to a very curious tale," said Lucretia. "I assure you it is absolutely true."

"When the College decided to install a stained glass window to commemorate Fisher and Venn, it asked a mathematical biologist, Dr. Anthony Edwards, to design it. And while he was doing that, he hit upon a beautiful solution to the problem of finding an n-set Venn diagram for arbitrarily large n. His sets [Figure 25] are, as Venn described, toothed, but more like cogwheels than combs."

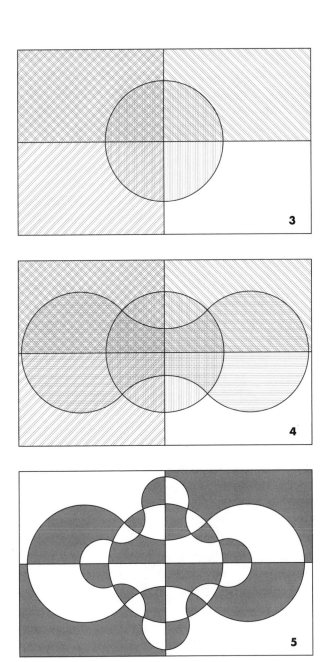

FIGURE **25** The Edwards-Venn cogwheels, which generalize to any number of sets. (*Continued on next page*)

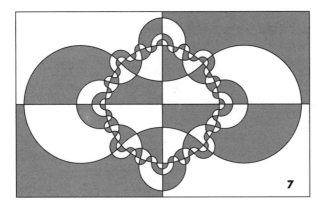

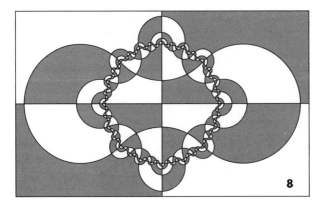

FIGURE **25**   (*Continued*)

"Cogwheels of the mind," said Jules. "That would make a terrific title for a book."

"He thereby solved a problem that had remained unanswered for over a century," Lucretia concluded.

"Golly!" said Jules.

"It's one of those discoveries that are very hard to make, but very easy to belittle," I said. "It's not famous like the Poincaré Conjecture or the Riemann Hypothesis."

"Those are famous?" asked Jules.

"Among mathematicians. Venn diagrams aren't exactly the mainstream of mathematical research anymore. But, important or not, they picked up an intellectual loose end that really did need tidying up. And the answer is simple and elegant, the essence of mathematics. I think it's extremely clever. It grows on you."

"And it does *relate* to the mainstream," said Lucretia. "It has applications to combinatorics."

"Not only that!" I said. This was getting exciting. "In keeping with the times, the diagram is a fractal. I mean, the picture for infinitely many sets has structure on all scales, and is defined by a recursive procedure. Thus the Edwards-Venn diagrams are, you know, a visual icon for the modern era."

"I think we'd better go home," said Lucretia suddenly. "Ian's starting to sound just like you, Jules."

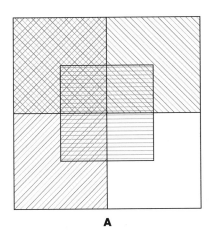
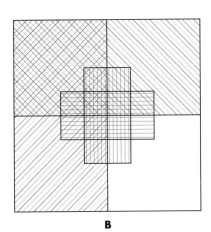

A   B

FIGURE **26** Lewis Carroll's alternative diagrams.

ANSWERS

1. Lewis Carroll's Venn diagrams are shown in Figure 26.

FURTHER READING

Dodgson, C. L. (Lewis Carroll). *Symbolic Logic.* London: Macmillan, 1896.

Clarence, Irving Lewis, and Cooper Harold Langford. History of symbolic logic. In *The World of Mathematics,* ed. James R. Newman, vol. 3. New York: Simon and Schuster, 1956.

Stewart, Ian, and David Tall. *The Foundations of Mathematics.* Oxford and New York: Oxford University Press, 1988.

Venn, John. On the diagrammatic and mechanical representation of propositions and reasonings. *Philosophical Magazine* (5th series) 10 (1880): 1–18.

———. *Symbolic Logic.* London: Macmillan: 1881.

How Many Goats in the Orchard?

●

●

●

●

●

"**B**arney!"

Farmer Quinn was not amused. "Barney, you'm let they danged goats into the orchard again!"

"Yes, Al. It be toime ter graze 'em."

Quinn sighed. "Not until the apples have been picked, Barney! They'm a bit late this year on account of the drought." It had been a dry summer in Weffolk. "You know what they gets up to, Barney! They danged goats be buttin' the trees to shake down the apples, and then they scoffs the lot! Get 'em out of there, now!" He threw his hat to the ground and jumped on it a few times to vent his anger. Then he walked around the orchard counting how many trees had lost their apples: there were 314 of them. He leaned on the top bar of the gate, chewing at a straw, lost in thought.

Eventually he came to a decision. "Well, may as well make the most of it, as ol' granddaddy Herbert would say. Barney! Fence off the trees as have lost their apples, an' put they goats to graze between 'em. But make sure you keeps they danged goats away from the rest of the crop!''

Barney hastened to obey, and for the rest of the afternoon all you could hear was the banging of hammers as a wire mesh fence was nailed to the trees, and the bleating of startled goats.

In mid-afternoon, Barney reappeared, bedraggled and sweating. "All finished, Al. Hard work, oi nailed that fence to a 'undred and noinety-noine trees! Oi counted as oi went. Oi'm just goin' to set the goats loose."

"Right. But make sure you don't let 'em overgraze the orchard."

"Eh?"

"Look, Barney, it be dead easy. The orchard be planted on a square grid, right?"

"Yes, Al."

"Good. Now each goat can graze exactly the amount of grass contained in one square cell of the grid. Got that? One square, one goat."

"Roight!"

"So all you need to do is count the number of squares enclosed by your fence, and that's 'ow many goats you can let in."

"Oh. Yes. Um. Al?"

"Yes, Barney?" said Algernon Quinn in exasperation.

"It b'ain't be made up of squares, Al. I 'ad to rig some of the fence kind of diagonal, loike."

"Then find its total area in units of one grid-square."

"Oi dunno 'ow to do that, Al."

"Ah, thee be thick as two short planks, Barney. Any danged fool can work out an area!"

"The fence be a bit of an odd shape, Al" (Figure 27). Algernon went to look. After more hat-stomping and a great deal of shouting he returned with Barney to the farmhouse and phoned up Jacob Staff, a surveyor from the nearby town of Upward-le-Mobile.

"Now that's what I call an *interesting* problem," said Jacob. "I could make a plan of the field for you, and then calculate the area by triangulation—"

"Foine!"

"—provided you're willing to pay the standard fee."

"Fee? What fee?"

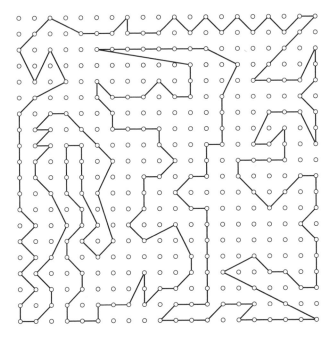

FIGURE **27**   How many goats can graze in the fenced-off part of the orchard? (Each goat needs the amount of grass in one unit cell of the square grid).

"A thousand pounds for the map, and another thousand for the triangulation."

"*Two thousand pounds?* Ol' granddaddy Herbert would turn in his grave. Except he b'ain't be dead yet. Then again, it moight well pop 'is clogs for 'im. Never heard such nonsense."

"Well, there is a cheaper way. It's an old surveyor's trick. You see, you've planted your orchard in what we mathematical types call a lattice."

"Nope, they be apples, not lettuce!"

Jacob hastened to clarify. "A lattice is a square grid, Algernon."

"Ah. Whoi didn't you say so the first toime?"

"I thought I did, but no matter. As it happens, you've also stretched the fence from tree to tree, so that it forms a lattice polygon. Now you've told me already that there are 314 trees on or within the polygon. Of those exactly 199 are on the boundary, because Barney nailed the fence to that number of trees. So there are 115 trees *inside* the fence.''

''Roight. Though what use all that be, oi can't imagine.''

''Oh, but it is! There's a remarkable formula, called Pick's Theorem. It was discovered in 1899 by G. Pick, and it lets you work out the area of a lattice polygon. All you need to know is the number of lattice points — trees — inside it, and the number on the boundary. It's extremely simple to use . . . and the charge is a mere fifty pounds.''

''Oh ar. Fifty. Fair enough, oi guess.'' There was a lengthy pause. ''Go on, then.''

''There's a minor snag. I can't remember the formula,'' said Jacob.

''Surveyor, thee be as much use as a teaspoon at muckspreadin' toime.''

''Don't be so hasty, Algernon! Merely *knowing* that there is a formula helps enormously.''

''Whereas *knowin'* there be an area b'aint' no bleedin' use?''

''That's different. We can always work out the formula, once we know there is one. Working out the area gets us back to the business of the two-thousand-pound fee.''

''Foire away then, Jake. But oi warns 'ee, we farmers b'ain't so hot at algebra an' geometry an' things loike that. Sheep, now, we be real good at sheep.''

''Woolly thinking is very dangerous in mathematics, Algernon. I think algebra will get us further than sheep, in this particular instance.''

''Loike as 'ow?''

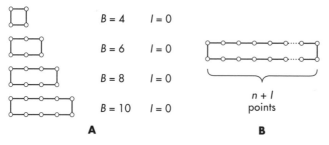

FIGURE 28   *A.* $1 \times n$ rectangles for $n = 1, 2, 3, 4$. *B.* The general $1 \times n$ rectangle.

"Boi—I mean by, considering special cases. I'll let A denote the area of the polygon, B the number of boundary points, and I the number of interior points. And I'll look at simple cases. For instance, a 1×1 square has $A = 1$, $B = 4$, and $I = 0$. Similarly a 1×2 rectangle has $A = 2$, $B = 6$, $I = 0$; a 1×3 rectangle has $A = 3$, $B = 8$, $I = 0$; and so on [Figure 28 A]. We can make a little table."

"Roight! Oi'll go get a saw."

"No, what I mean is—look, like *this.*"

Size	A	B	I
1×1	1	4	0
1×2	2	6	0
1×3	3	8	0
1×4	4	10	0

"Oh ar."

"So—can you spot any pattern in the numbers?"

"Yur. Column I don't got nuffin' in it."

"I agree that I is always zero, Algernon, but that's not quite the pattern I had in mind, and in fact it worries me a bit because it suggests that I haven't chosen sufficiently general polygons. However, I can see some sort of pattern. The value of B is *nearly* twice that of A. In fact, we have

$$B = 2A + 2$$

in each case. Solving for the area, we get

$$A = \tfrac{1}{2}B - 1$$

The area is half the number of boundary points, less one."

"Good thinkin', squire!"

"In fact, we can prove it for a general $1 \times n$ rectangle [Figure 28 B]. The area's easy: it's n units, where each unit is the area of a single square lattice cell. There are $n + 1$ lattice points along the top row, and $n + 1$ along the bottom, making $2n + 2$ altogether."

"Oh, ar. We farmers be good with 'ens. Two 'en plus two, that be four 'ens altogether. Unless the two be cocks—"

"Algernon, n, not *hen*! It's a symbol for an arbitrary number!"

"Oh."

"No, n, not o!"

"Ar."

It could have got very confusing, but Jacob eventually realized that Quinn was merely commenting, not suggesting new symbols, and stopped correcting him. "So $A = n$, $B = 2n + 2$, $I = 0$, and the formula holds," he concluded triumphantly.

"Oh ar. You'm a genius, Jake."

"To check, let's try a 2×2 square. Now $A = 4$, $B = 8$, $I = 1$, and $\frac{1}{2}B - 1 = 4 - 1 = 3$. Bother! It's not working!"

"Ar, but now the number of interior points b'ain't nuffin' no more."

"That's a stupid thing to—I mean, good grief, that's absolutely right, Algernon! Hmmm . . . let's try some $2 \times n$ rectangles, and see how A and $\frac{1}{2}B - 1$ compare . . ."

Size	A	B	I	$\frac{1}{2}B - 1$	$A - (\frac{1}{2}B - 1)$
2×1	2	6	0	2	0
2×2	4	8	1	3	1
2×3	6	10	2	4	2
2×4	8	12	3	5	3
2×5	10	14	4	6	4
2×6	12	16	5	7	5

"Aha! *Now* I can see a pattern!"

"Now *any* danged fool can see a pattern! Plain as a piglet's snout, it be!"

"It certainly is, Algernon. The last column is the same as column I. So that suggests that $A - (\frac{1}{2}B - 1) = I$, which can be rewritten to give:

PICK'S THEOREM

For any lattice polygon, the area A *is given in terms of the number of boundary points* B *and interior points* I *by the formula*

$$A = \tfrac{1}{2}B + I - 1$$

Isn't that beautiful!''

"It sure be, Jake. One thousand, noine hundred and fifty pounds bootiful.''

"Mmpph. But we haven't proved it yet.''

"Eh? Wot we just bin doing', then?''

"Motivation and conjecture, Algernon, nothing more. Clearly the next step is to try it out on other examples.''

PROBLEM ❶

Try out Pick's Theorem on the lattice polygons in Figure 29.

"That's good enough for me, squire!'' said Quinn, after a certain amount of discussion. "Let's see, for the orchard, the number of boundary trees be — ''

"Not so fast, not so fast! Hold your horses!''

"I can see you b'ain't a countryman, Jake. Them be goats. You can tell by the little 'orns an' the big smell.''

"Quite. No, Algernon, we still have to *prove* that the theorem is true.''

"But we'm got oodles of evidence!''

"We thought we had oodles of evidence that $A = \tfrac{1}{2}B - 1$, but we were wrong. How do you know we're not still wrong?''

"I make a h'eduficated guess!''

"Mmmph. No, despite your erudition, I insist on a proof. Let's think about the $1 \times n$ case again. Suppose we add a single square to one end

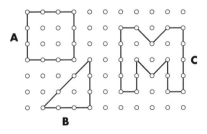

FIGURE **29**   Some lattice polygons to experiment with.

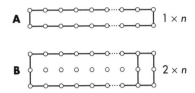

FIGURE **30**   Adding a square to one end of a rectangle.

[Figure 30 *A*]. Then *A* increases by 1, *B* increases by 2, *I* stays the same (zero). So $\frac{1}{2}B + I - 1$ increases by 1, which is just like *A*. And in the 2 × *n* case [Figure 30 *B*] we add a 2 × 1 block, so that *A* increases by 2, *B* increases by 2, *I* increases by 1, and $\frac{1}{2}B + I - 1$ increases by 2, just like *A*. Aha! I've got an idea! *If you join two lattice polygons together, the expression* $\frac{1}{2}B + I - 1$ *just adds up.* It works, too: see the box [to the right]."

"It certainly is. You see, it tells us the whenever Pick's Theorem is true for two lattice polygons Q and R, then it's true for their join P as well."

"Whoi?"

"Because the areas add up just the same as the picks do."

"Oh! Roight! Good on yer, squire, I sees that. Clear as mud, it be!"

"Now that gives us an excellent proof strategy. We can prove the theorem for a general lattice polygon provided we can break it up into

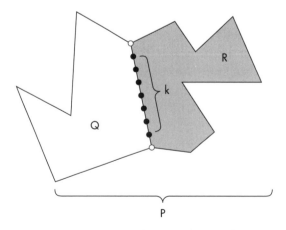

FIGURE **31**   Joining two lattice polygons along a common boundary.

$\frac{1}{2}B + I - 1$ IS ADDITIVE

Consider two general lattice polygons Q and R, joined together as in Figure 31. Let P be the polygon that results from joining them. Suppose the common boundary has k interior points on it (solid dots), plus two at the ends (open dots). For the lattice polygon P define

$$\text{pick}(P) = \tfrac{1}{2}B_P + I_P - 1$$

where B_P is the number of boundary points for P, and I_P is the number of interior points; similarly for pick(Q) and pick(R). I claim that the expression *pick* is additive, that is, if P is obtained by joining Q and R together then pick(P) = pick(Q) + pick(R). Now

$$\text{pick}(P) = \tfrac{1}{2}B_P + I_P - 1,$$

and we can see from Figure 31 that

$$B_P = B_Q + B_R - 2k - 2$$

That's because the k solid points used to be boundary points for both Q and R, but now they're interior points for P so we have to subtract $2k$; and the two open boundary points for P used to be boundary points for both Q and R, so they originally counted as 4, but now only as 2, so we subtract a further 2. In the same way

$$I_P = I_Q + I_R + k$$

because anything that was an interior point for Q or R remains an interior point for P, but we get a further k solid interior points.
So

$$\text{pick}(P) = \tfrac{1}{2}(B_Q + B_R - 2k - 2) + (I_Q + I_R + k) - 1,$$

which is equal to

$$[\tfrac{1}{2}B_Q + I_Q - 1] + [\tfrac{1}{2}B_R + I_R - 1]$$

and that's just

$$\text{pick}(Q) + \text{pick}(R)$$

as claimed.

suitable pieces and prove the theorem for each piece. So — what sort of piece should we pick — I mean, choose?"

"Squares?"

"No, we want to handle completely general polygons with sloping sides."

"Ar. Troiangles?"

"You're got it! Look, *every* polygon — lattice or not — can be cut up into triangles" (Figure 32 A).

"Roight."

"Now, we can express any triangle in terms of right-angled triangles and rectangles. For example, take this one." Jake quickly drew Figure 32 B. "Call the big triangle T and the pieces U, V, W, X. Then

$$\text{pick(T)} = \text{pick(U)} + \text{pick(V)} + \text{pick(W)} + \text{pick(X)}$$

and also

$$\text{area(T)} = \text{area(U)} + \text{area(V)} + \text{area(W)} + \text{area(X)}$$

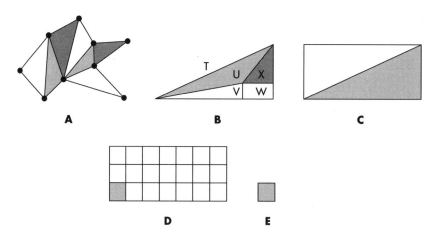

FIGURE **32** Steps in the proof of Pick's Theorem. Complicated cases are reduced to simpler ones. *A.* If the theorem is true for triangles, then it is true for all lattice polygons. *B.* If it is true for right-angled triangles, then it is true for arbitrary triangles. *C.* If it is true for rectangles, then it is true for right-angled triangles. *D.* If it is true for 1×1 squares, then it is true for rectangles. *E.* The only case we need to check!

Now, if we know that pick = area for all rectangles and right-angled triangles, then we know that this holds for T, V, W, and X. Comparing the two equations we find it also holds for U, the triangle we want.

"Next, observe that right-angled triangles are half rectangles [Figure 32 C], and again the picks and areas behave the same way: each is half that for the rectangle. So the truth of Pick's Theorem for rectangles implies its truth for right-angled triangles. But rectangles are built up from unit squares [Figure 32 D]. So finally it all boils down to checking Pick's Theorem for the unit square [Figure 32 E] . . . which of course we've already done in the first table!"

PROBLEM ❷

If you're still not convinced, try some more experiments (Figure 33). Find their areas without using Pick's Theorem, then using it, and verify that the answers agree.

PROBLEM ❸

Can you generalize Pick's Theorem to polyhedra whose vertices lie on a three-dimensional lattice?

"Now, Algernon: you can decide how many goats you can let into the orchard."

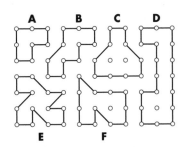

FIGURE 33   Try out Pick's Thorem on these lattice polygons.

FIGURE **34**   What's gone wrong now?

"Oh ar, roight. Lessee—we'm got $B = 199$, an' $I = 115$. So the area should be $A = \frac{1}{2}B + I - 1 = 99\frac{1}{2} + 115 - 1 = 213\frac{1}{2}$. Roight!" He turned to his farmhand. "Barney! You get two 'undred an' thirteen an' a 'alf goats into that there orchard roight away!"

"Algernon, 'ow do oi get 'old of 'alf a goat?"

"Eh? Oh . . . get the big knife from the—"

"I suggest using a kid, Algernon. A goat that's been sliced in half won't eat any grass at all."

"True, true, you be a deep sort, Jake, an' no mistake." Algernon nodded sagely. He bent to wipe something off his boots, and stomped off to where eight concrete slabs had been arranged in a square (Figure 34) to keep one corner of the farmyard free from mud and other insalubrious substances. "That be interestin'," he said. "Another lattice polygon. Moight as well give old Picky a final fling, eh? Lemme see . . . we'm got $B = 16$, $I = 0$, so the area ought to be $\frac{1}{2}B + I - 1 = 8 + 0 - 1 = 7$. But it b'ain't, it be 8! Jacob, there be a small problem here!"

PROBLEM ❹

What's gone wrong, and why? What assumption was hidden in our alleged "proof" of the formula? How can you modify the formula to apply to this case too?

ANSWERS

1. "A 3×3 square [Figure $29A$], for instance," said Al Quinn, has $A = 9$, $B = 12$, $I = 4$. So $\frac{1}{2}B + I - 1 = 6 + 4 - 1 = 9$, which is right."

"Yes," said Staff, "but the formula mustn't just work for rectangles, that's much too simple a case. What about a triangle [Figure 29 B], half a 3 × 3 square?"

"Then," replied Quinn, "the area is $A = 4\frac{1}{2}$ because it's half 9; and $B = 9$, $I = 1$. Now $\frac{1}{2}B + I - 1 = 4\frac{1}{2} + 1 - 1 = 4\frac{1}{2}$, and it works again! And the M-shaped object [Figure 29 C]," he continued, "has $B = 20$, $I = 3$, and that should mean $A = 10 + 3 - 1 = 12$."

2. Here's a table to show what happens:

Shape	I	B	$\frac{1}{2}B+I-1$	Area
A	0	8	3	3
B	0	11	$4\frac{1}{2}$	$4\frac{1}{2}$
C	2	10	6	6
D	1	18	9	9
E	0	14	6	6
F	1	12	6	6

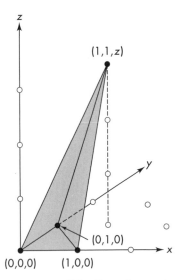

FIGURE **35** John Reeve's counterexample to a simple three-dimensional Pick's Theorem.

3. There is no formula for the volume of a lattice polyhedron in terms of the numbers of points in its interior or on its boundary. This was shown by John Reeve in 1957. A pyramid with corners at (0,0,0), (1,0,0), (0,1,0), and (1,1,z) always has 4 boundary points and 0 interior points, but its volume is $\frac{1}{6}z$ which varies with z (Figure 35). However, Reeve also shows that if we introduce a second lattice, for example one in which half-integer coordinates are allowed, then there exists a formula that works for all convex lattice polyhedra. The formula is too complicated to give here: Reeve's paper is listed in *Further Reading*.

4. The problem is the hole, which causes the proof of Pick's Theorem that we gave to break down. The proof assumes that the two parts Q and R separate when P is cut along a single common line. To modify the formula, you have to add the number of holes. This modification is also discussed in Reeve's paper, and forms the starting point for his three-dimensional generalization. He also makes conjectures about higher dimensions.

FURTHER READING

Coxeter, H. S. M. *Introduction to Geometry.* New York: John Wiley & Sons, 1969.

Pick, G. Geometrisches zur Zahlenlehre. *Zeitschrift der Vereines "Lotos."* Prague, 1899.

Reeve, J. E. On the volume of lattice polyhedra. *Proceedings of the London Mathematical Society* (3d series) 7 (1957):378–95.

Steinhaus, H. *Mathematical Snapshots.* Oxford and New York: Oxford University Press, 1950.

Passage to

Pentagonia

———

The book, *A Survey of Matrix Theory*, was black and weighty. The brochure, in contrast, was slim and glossy, with a picture of a sun-drenched beach, palm trees, and a bronzed beauty in a bikini.

In the contest for my attention, the brochure won.

It was accompanied by a computer-printed form letter, which began:

Dear Mr. Sheward,
    This is your chance for the *holiday* of a LIFETIME!!!

and continued in that sort of typographer's nightmare for two more pages. *Hmmmph*, I thought, *They can't even get my name right.* I put the holiday brochure down again and picked up the weighty tome. But outside a gale was howling, hailstones drummed on the roof, and on television the weathermen were forecasting earthquakes . . .

Actually, the place didn't look too bad. And the fare was pretty cheap, too. Special winter rate, only £120 for a whole week in—

In where? Where on earth was Pentagonia? The brochure said it was in the Budgerigar Isles, five hundred miles northwest of Ouagadougou, off the coast of Upper Volta.

It sounded exotic.

I read on.

The island paradise of Pentagonia, the brochure explained, is famous for its scenic railway. Built by occupying Italian troops in the 1890s, the railway winds picturesquely round the beautiful coast. It visits the five towns of Abbindolare (the capital), Bancarotta, Canzonatura, Dappoco, and Esoso. The track forms a complete loop, and trains run in both directions (Figure 36). One of the enticements on offer was a free pass, for eight trips on the train, each between one town and the next.

The beaches and hotels looked too good to be true, so they probably were. Too good to be true, that is. I picked up *A Survey of Matrix Theory* again and started to wade through its turgid pages.

I must have dozed off. And I had a very strange dream.

I stood in front of a huge neon sign: WELCOME TO ABBINDO-LARE INTERCONTINENTAL AIRPORT. In my hand I clutched the

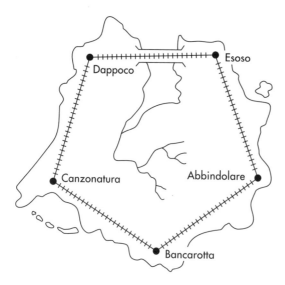

FIGURE **36** The island paradise of Pentagonia and its scenic railway.

free railway pass. *A circular tour round the coast,* I thought, *takes five trips.* Three trips out of the eight left over. Hmmm . . . don't want to waste them. I could visit Bancarotta again, there and back — no, that still leaves one trip. Can't leave Abbindolare on a one-way journey; need to get back to catch the plane home.

It's a fraud, I thought. *You can't use all eight tickets. What a sneaky trick!*

I quickly realized my mistake. There are many different ways to make a tour of eight trips, starting and ending at Abbindolare. For instance, representing the towns by their initials, ABABABABA. That might make for a boring holiday, though. ABCDEDCBA would be better. Or ABCBAEDEA, with the advantage of a visit to Abbindolare, the capital, in the middle.

PROBLEM ❶

In how many different *ways can I make a tour of exactly eight trips, starting and ending at Abbindolare?*

I sat in a pink wastebasket in the Abbindolare airport — as I said, it was a strange dream — and thought. Each trip must be either clockwise or counterclockwise. In order to return to the starting point in eight trips, there must be four of each. So the number of tours is the number of ways of dividing eight trips into four clockwise and four counterclockwise. But when the four clockwise trips have been chosen, the others are automatically counterclockwise. In other words, the number I want is the number of different ways to choose four objects out of eight. This can be calculated; the answer is 70.

However, that isn't the best way to tackle this kind of problem. For example, suppose there were a sixth town — Fallacia, say — in the loop. At first sight you might think the answer is also 70 tours, for the same reason. But in fact there are now 86 different tours. The reason is that with six towns in the circuit, four trips clockwise and four counterclockwise is *not* the only way to get back to Abbindolare. It could also be seven trips clockwise and one counterclockwise, or seven counterclockwise and one clockwise. Each of these accounts for an extra eight tours, providing the missing 16.

What happens with other numbers of towns, or other numbers of trips? In particular, how many different tours of n trips can I make on Pentagonia Rail, returning to my starting point? Let's try to find out.

I began writing on the sandy floor of the arrivals lounge, while a green toucan sat on a lamppost and winked at me. I drew up a table, like this:

n	A	B	C	D	E
0	1	0	0	0	0
1	0	1	0	0	1
2	2	0	1	1	0
3	0	3	1	1	3
4	6	1	4	4	1
5	2	10	5	5	10
6	20	7	15	15	7
7	14	35	22	22	35
8	70	36	57	57	36

The nth row shows the number of tours of length n that start at A and end in the given column. Let me explain how the calculation goes.

A typical case is row 6, column C, with the entry 15. In how many ways can I get from A to C in six trips? Well, the sixth trip must end at C, so it is either BC or DC, because B and D are the only towns that connect to C. So I have to follow a five-trip journey from A to B, and then go to C; or I have to follow a five-trip journey from A to D, and then go to C. In other words, the number I want is the sum of two numbers in row 5, namely those in columns B and D. This is $10 + 5 = 15$.

The same reasoning gives a rule for generating the table. *Each number is the sum of the numbers to its left and right in the row above.* (The right-hand end of the row must be considered as "wrapping around" to be adjacent to the left-hand end.)

To get started, we need to know row 0. So suppose I start at A and make 0 trips. Obviously I can only get to A. So the A entry in row 0 is 1, and all the others are 0. But now I can use the rule above to calculate row 1, then row 2, and so on. Note that row 8, column A, is 70, a finding that confirms my previous calculation. The same row shows that there are exactly 36 different ways to get from A to B or E in eight trips, and 57 ways to get to C or D.

This is an effective method for calculating the numbers of tours that take a small number of trips. But suppose I wanted to know the number of tours with trips? 100? It would take forever to work out the table. Is there some general pattern to save all that work?

I extended the table to rows 9, 10, 11, . . . , desperately searching for a pattern in the numbers. My calculations spread across half of Abbindolare Intercontinental. The toucan started to eat my luggage. It bit a piece off *A Survey of Matrix Algebra* but had to spit it out. Indigestible.

At that point, I woke up. The book was lying on the floor. I looked at the page it was open to—and I realized that there is a connection between the problem of making tours on Pentagonia Rail and matrix algebra.

A matrix is a square (or rectangular) array of numbers. Matrices were invented about a century ago, by the English mathematician Arthur Cayley. "This idea cannot possibly have any applications," said Cayley. He could hardly have been more wrong: today, engineers, economists, physicists, biologists, and statisticians all use matrix algebra in their daily work.

There is an algebra of matrices. You can add them together, or multiply them. The rule for adding is straightforward: to add two matrices M and N, just add the entries in corresponding positions. But we don't need that here. What we need is the rule for *multiplying M and N* to get $M \times N$. This is much more peculiar: see Figure 37. Using it, you can also calculate the powers $M^2 = M \times M$, $M^3 = M \times M \times M$, and so on, of a square matrix M. Powers of matrices hold the key to the Pentagonia Problem.

The Pentagonia Rail network can be described by its incidence matrix. This is a table of numbers that indicates what items are to be joined, in this case showing which towns are joined by railway lines. If towns X and Y have a line between them, you write 1 in row X and column Y of the incidence matrix; otherwise you write 0. So the incidence matrix for Pentagonia Rail looks like this:

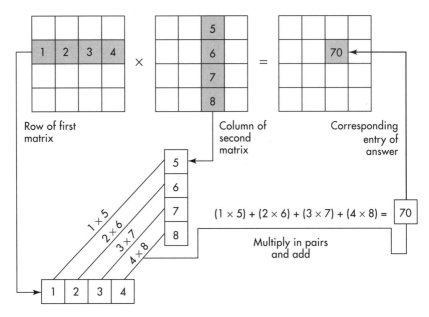

FIGURE **37**   The rule for matrix multiplication.

	A	B	C	D	E
A	0	1	0	0	1
B	1	0	1	0	0
C	0	1	0	1	0
D	0	0	1	0	1
E	1	0	0	1	0

For example, towns C and D are connected, so there is a 1 in row C, column D; towns E and B are not, so there is a 0 in row E, column B.

Any system of towns and railways lines—abstractly, any graph consisting of dots joined by lines—has an incidence matrix, defined this way (Figure 38). Since C is connected to D, D is also connected to C, so there also has to be a 1 in row D, column C. In other words, the incidence matrix must be symmetric, that is, each half the same as its reflection across the diagonal line sloping from top left to bottom right.

Suppose M is the incidence matrix for Pentagonia Rail. Can you work out M^2, M^3, and so on? Some of my answers are in Figure 39, worked out using the multiplication rule described above. Compare the numbers with those in the table, and you'll see a striking pattern.

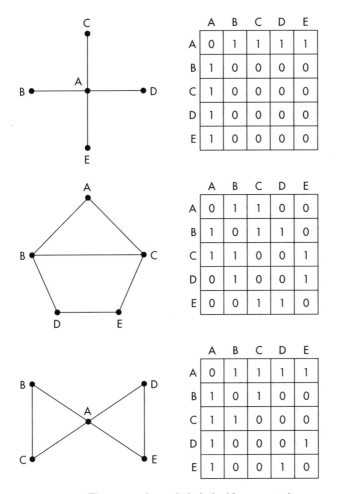

FIGURE **38**   Three graphs and their incidence matrices.

Namely, the top row of M^n is just the nth entry in the table. The other rows of M^n are similar, but successively displaced one space to the right.

This is no coincidence. For *any* graph, the entry in row X and column Y of the nth power of the incidence matrix is always equal to the number of distinct tours, consisting of n trips, that start at town X and end at Y. This follows from the rule of matrix multiplication, and the fact that any tour of length n from X to Y can be broken up as a tour of

1st power of matrix M

0	1	0	0	1
1	0	1	0	0
0	1	0	1	0
0	0	1	0	1
1	0	0	1	0

5th power of matrix M^5

2	10	5	5	10
10	2	10	5	5
5	10	2	10	5
5	5	10	2	10
10	5	5	10	2

2nd power of matrix M^2

2	0	1	1	0
0	2	0	1	1
1	0	2	0	1
1	1	0	2	0
0	1	1	0	2

6th power of matrix M^6

20	7	15	15	7
7	20	7	15	15
15	7	20	7	15
15	15	7	20	7
7	15	15	7	20

3rd power of matrix M^3

0	3	1	1	3
3	0	3	1	1
1	3	0	3	1
1	1	3	0	3
3	1	1	3	0

7th power of matrix M^7

14	35	22	22	35
35	14	35	22	22
22	35	14	35	22
22	22	35	14	35
35	22	22	35	14

4th power of matrix M^4

6	1	4	4	1
1	6	1	4	4
4	1	6	1	4
4	4	1	6	1
1	4	4	1	6

8th power of matrix M^8

70	36	57	57	36
36	70	36	57	57
57	36	70	36	57
57	57	36	70	36
36	57	57	36	70

FIGURE **39**   Powers of the Pentagonian incidence matrix.

length $n - 1$ from X to some Z, followed by a tour of length 1 from Z to Y.

You can use this result to calculate the number of tours of given length between two points in any graph. For example, Figure 40 shows

252	1	210	10	120	45	45	120	10	210	1
1	252	1	210	10	120	45	45	120	10	210
210	1	252	1	210	10	120	45	45	120	10
10	210	1	252	1	210	10	120	45	45	120
120	10	210	1	252	1	210	10	120	45	45
45	120	10	210	1	252	1	210	10	120	45
45	45	120	10	210	1	252	1	210	10	120
120	45	45	120	10	210	1	252	1	210	10
10	120	45	45	120	10	210	1	252	1	210
210	10	120	45	45	120	10	210	1	252	1
1	210	10	120	45	45	120	10	210	1	252

FIGURE **40**   The tenth power of the incidence matrix for a circuit of 11 towns.

the tenth power of the incidence matrix for 11 towns in a circular loop. It shows that there are 252 ways to complete a ten-trip round tour, but only one way to go from a given town to its nearest neighbor in ten trips! Why is this?

What about general patterns? Let's start with a simpler problem: powers of the incidence matrix for a four-town circular loop. Here's what you get:

First power

0	1	0	1
1	0	1	0
0	1	0	1
1	0	1	0

Second power

2	0	2	0
0	2	0	2
2	0	2	0
0	2	0	2

Third power

0	4	0	4
4	0	4	0
0	4	0	4
4	0	4	0

Fourth power

$$
\begin{array}{cccc}
8 & 0 & 8 & 0 \\
0 & 8 & 0 & 8 \\
8 & 0 & 8 & 0 \\
0 & 8 & 0 & 8
\end{array}
$$

In this case the pattern could hardly be more striking. The numbers form a kind of checkerboard and double each time, while the positions of the 0's alternate. The number of tours from a town to itself on a four-town circuit is

$$
\begin{array}{ll}
0 & (n \text{ odd}) \\
2^{n-1} & (n \text{ even})
\end{array}
$$

A more compact way to write this is

$$
\frac{1}{4}(2^n + (-2)^n)
$$

Here the distinction between even and odd n is buried in the minus sign. Is there a similar formula for tours by Pentagonia Rail to five towns instead of four? Indeed there is. Remarkably, it involves the golden number

$$
\phi = \frac{1 + \sqrt{5}}{2} = 1.618034. \ . \ .
$$

which has an uncanny affinity with pentagons. The formula is:

$$
\frac{1}{5}\{2^n + 2(\phi - 1)^n + 2(-\phi)^n\}
$$

The proof involves matrix algegra but is too advanced to give here. When you first look at the formula, it isn't at all obvious that its value is a whole number. But it is!

In fact, there is a general formula for loops of m towns. I'll express it geometrically. Draw a circle of radius 2, and draw a regular m-sided polygon inside it, as in Figure 41. Project the vertices on to the axis as shown, and let the resulting distances from the center be $d_1, d_2, \ . \ . \ .,$ d_m.

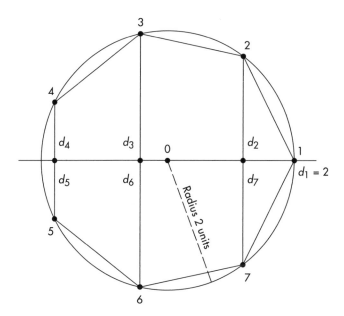

FIGURE 41 A geometric prescription for the number of n-trip round tours on a circuit of m towns. Draw a regular m-sided polygon in a circle of radius 2, project to get distances d_k, and find the average of the nth powers of the d_k.

PROBLEM ❷

Using trigonometry, find a formula for d_k.

Count distances to the left as negative. Then the number of round tours of n trips is

$$\frac{1}{m}(d_1^n + d_2^n + \ldots + d_m^n)$$

That is, it is *the average of the nth powers of the projected distances.* This is an even more remarkable result. It is even less obvious that the sum of these nth powers is a whole number. But the formula says it must be, for *any* n; and furthermore, that whole number is divisible by m.

For example, suppose $m = 7$. The projected distances are:

$$d_1 = 2 \cos (360 \times 0/7) = 2$$
$$d_2 = 2 \cos (360 \times 1/7) = 1.24698$$
$$d_3 = 2 \cos (360 \times 2/7) = -0.44504$$

$$d_4 = 2 \cos (360 \times 3/7) = -1.80194$$
$$d_5 = 2 \cos (360 \times 4/7) = -1.80194$$
$$d_6 = 2 \cos (360 \times 5/7) = -0.44504$$
$$d_7 = 2 \cos (360 \times 6/7) = 1.24698$$

So the number of tours of length n, starting and ending at the same town, is one seventh of the sum of the nth powers of these seven numbers. As a check, on a computer, the first few values are:

Exact	Computed
0	0.00000018
2	2
0	0.00000040
6	6.000001
0	0.00000027
20	20
2	1.999996
70	70.00002
18	17.99999
252	252

I repeat, the formula is *exact*: the tiny discrepancies in the table are due to rounding errors in the computer arithmetic.

To get my formula for Pentagonia Rail, all you need to know is that when $m = 5$ we have $d_1 = 2$, $d_2 = d_5 = \phi - 1$, $d_3 = d_4 = -\phi$. You can recover the formula for $m = 4$, too, by observing that here $d_1 = 2$, $d_2 = d_4 = 0$, $d_3 = -2$.

We started with a problem about railways. Now we have journeyed, via graph theory and matrix theory, to a result in trigonometry. Isn't mathematics amazing?

PROBLEM ❸

How many 50-trip tours are there starting and ending at Abbindolare?

PROBLEM ④

If an extra railway line is built between Canzonatura and Esoso, how many eight-trip tours are there starting and ending at Abbindolare?

I relaxed euphorically in my chair, feeling very satisfied by my train of thought. *I will visit Pentagonia. I'll experience the scenic tour at first hand. Where's that atlas? Where* exactly are *the Budgerigar Isles?*

I couldn't find them.

But I did discover a number of interesting things.

A hundred miles northwest of Ouagadougou puts you in the Sahara Desert. Upper Volta is an inland state: it has no coast. And in Italian, "Abbindolare," "Bancarotta," "Canzonatura," "Dappoco," and "Esoso" mean "cheat," "bankrupt," "hoax," "useless," and "detestable."

I think I'll stick to matrix theory.

ANSWERS

1. There are precisely 70 eight-trip tours of Pentagonia:

ABCDEDCBA	ABCDCDCBA	ABCDCBCBA
ABCDCBABA	ABCDCBAEA	ABCBCDCBA
ABCBCBCBA	ABCBCBABA	ABCBCBAEA
ABCBABCBA	ABCBABABA	ABCBABAEA
ABCBAEABA	ABCBAEAEA	ABCBAEDEA
ABABCDCBA	ABABCBCBA	ABABCBABA
ABABCBAEA	ABABABCBA	ABABABABA
ABABABAEA	ABABAEABA	ABABAEAEA
ABABAEDEA	ABAEABCBA	ABAEABABA
ABAEABAEA	ABAEAEABA	ABAEAEAEA
ABAEAEDEA	ABAEDEABA	ABAEDEAEA
ABAEDEDEA	ABAEDCDEA	AEABCDCBA
AEABCBCBA	AEABCBABA	AEABCBAEA
AEABABCBA	AEABABABA	AEABABAEA
AEABAEABA	AEABAEAEA	AEABAEDEA
AEAEABCBA	AEAEABABA	AEAEABAEA
AEAEAEABA	AEAEAEAEA	AEAEAEDEA
AEAEDEABA	AEAEDEAEA	AEAEDEDEA

AEAEDCDEA	AEDEABCBA	AEDEABABA
AEDEABAEA	AEDEAEABA	AEDEAEAEA
AEDEAEDEA	AEDEDEABA	AEDEDEAEA
AEDEDEDEA	AEDEDCDEA	AEDCDEABA
AEDCDEAEA	AEDCDEDEA	AEDCDCDEA
AEDCBCDEA		

2.
$$d_{k+1} = 2 \cos \frac{360k}{m}$$

3. There are precisely 225,191,238,869,774 tours of length 50 from Abbindolare to itself.

4. If an extra railway line is built between Canzonatura and Esoso, there are 189 eight-trip tours starting and ending at Abbindolare.

FURTHER READING

Harary, F. *Graph Theory*. Reading, Mass.: Addison-Wesley, 1969.

Read, Ronald. The graph theorists who count — and what they count. In *The Mathematical Gardner*, ed. David A. Klarner. Boston: Prindle, Weber and Schmidt, 1981.

Tutte, W. T. *Graph Theory*. Reading, Mass.: Addison-Wesley, 1984.

Wilson, Robin J. *Introduction to Graph Theory*. Harlow, England: Longman, 1985; New York: John Wiley & Sons, 1985.

Knights of the Flat Torus

"It is finished, your majesty."

King Arthur leaned over the edge of the balcony to inspect the Uther Pendragon Jousting Arena and Fishmarket. *Nothing like a good 'ournament 'o keep the people's minds off food shor'ages,* he thought to himself. It is little known that Arthur suffered a speech impediment and could not pronounce the letter t. Even his thoughts were afflicted. The smile suddenly faded from his bearded visage, as he realized that this arena was, indeed, *nothing* like a good tournament.

"Merlin!"

"Yes, my liege?"

"Is wondrous curious, this jous'ing-ground. Where will the chargers gallop?"

"On yonder field, Your Majesty, with the checkered pattern."

"But Merlin, i's *square*. And far 'oo shor' for the chargers 'o work up sufficien' speed.

"Ah. Your majesty is, as ever, correct. But by my magic have I fashioned a new manner of beast. Behold!" There was a puff of green smoke and a strange form appeared.

"I' faith, Merlin, i' resembles nothing more than a horse's head!"

"'Tis a new concept, my liege. I call it a knight."

"Sir Laughalo' will have thee slain for such a calumny."

"Of course, 'tis not a *real* knight. Your majesty will understand that 'tis more in the manner of a jest than a joust. I have fashioned a game, where war can be waged by wooden pieces." There was more green smoke and ranks of carved statues popped into view along two edges of the field.

"These shapes be passing odd, Merlin. Save yonder one resembling a castle."

"'Tis a rook, my liege."

"Merlin, if thou think'st tha' be a rook, I am glad thou'rt no' in charge of scaring the crows. What is i' called, this game of thine?"

"I call it chest, your majesty. Because the pieces, when not in use, reside in a wooden casket."

"Ches'." The king rolled the name round his tongue a few times. He liked the sound of it. Merlin sighed, and mentally renamed his brainchild. He hadn't yet had time to work out the rules, but he *had* decided how each piece should move. Arthur frowned as he discovered that the king could move only one square at a time. Especially when he was told that the queen could move as far as she liked in any direction.

"My liege, a single move is more dignified, more in keeping with kingly bearing," said Merlin firmly. "It is the duty of the other pieces to protect their king."

Arthur was especially delighted with the knight, though sad to discover that each king owned only two. "Whom shall I choose?" he inquired in anguish. "Sir Laughalo', of course, otherwise Gwinny will scold me for a month. But should the second be Sir Garrulous or Sir Belvedere? Whichever I choose, the other will sulk . . ."

Of the six different types of ches'piece, the knight stands out as an oddity. Unlike all other pieces, its move is a jump: it travels from its initial position to its final resting place without passing through any intermediate squares. The move itself is unusual, a dog-legged motion two squares horizontally or vertically followed by one square at right angles. "Like Sir Golaime," said the King, referring to a portly knight who had suffered a near-miss from a dragon and walked with a pronounced limp.

"And what are the rules of thy game of ches'?" asked Arthur. Arthur liked rules, being a ruler himself. As long as he didn't have to obey them, of course.

Merlin hadn't actually got round to deciding that yet, and he was forced to improvise. "They go on a — er, a *quest*, Your Majesty. The knight must compose his moves in such a manner as to visit every square of the ches'board exactly once."

This is of course a famous ches' — sorry, chess — puzzle. Its history, in our own world, is littered with the names of famous mathematicians. The earliest recorded solution is that of Abraham De Moivre, better known for his theorem about complex numbers. In De Moivre's solution, the knight ends his quest (or, as we say in our less colorful language, his tour) on a square that is not one move away from the starting square. Adrien-Marie Legendre improved on this, and found a solution in which the first and last squares are a single move apart, so that the tour closes up on itself into a single loop of 64 knight's moves. Such a tour is said to be re-entrant. Not to be outdone, Leonhard Euler — the most prolific mathematician of all time, who did a lot of his work when he was blind — found a re-entrant tour that visits two halves of the board in turn (Figure 42).

Using a spell for transference between future and past, Merlin caused the three mathematical giants to materialize before their eyes and re-create their solutions for the king's delectation. Then a new thought occurred to him. For which sizes of square ches'board is a

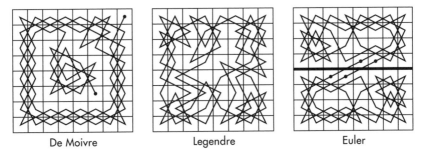

| De Moivre | Legendre | Euler |

FIGURE **42** Three tours by famous names.

knight's quest possible? With a wave of his wand and an appropriate incantation, the field of squares in the Uther Pendragon Jousting Arena and Fishmarket changed size.

Sir Laughalot came bounding in, a filmy pink favor tucked into his belt, waving his sword Exorbitant. "My liege, I bring urgent tidings of great portent—"

"Shu' up, Laughalo'. Merlin is explaining all abou' a *fascina'ing* new game of mine."

"But, your majesty—"

Arthur silenced him with a waggle of a finger. Laughalot looked uncomfortable but held his peace.

Merlin explained that knight's quests can be completed on boards of size 5 or greater (Figure 43). The quests shown on the 5 × 5 and 7 × 7 board are not re-entrant.

PROBLEM ❶

Can you see why it is impossible to find re-entrant quests on boards of these sizes? Is there a re-entrant quest on a board sized 999 × 999?

PROBLEM ❷

If you can answer Problem 1, can you decide whether or not a quest exists on the irregular octagonal board shown in Figure 44?

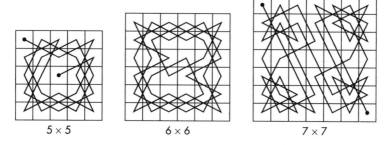

| 5 × 5 | 6 × 6 | 7 × 7 |

FIGURE **43**   Knight's quests for small squares.

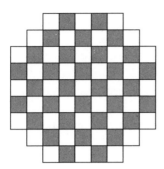

FIGURE **44**   Does a knight's quest exist?

"Merlin, thou hast made an omission."

"Your majesty?"

"A quest on a 4 × 4 board."

"My liege, no quest is shown for the 4 × 4 board, for a simple reason: it does not exist."

"Hmmph. How canst thou be sure?"

"I have tried many times, your majesty, but never succeeded."

"Thou hast also tried many times producing a maiden from thin air, and never succeeded," said Sir Laughalot dryly. "But that is no proof of the non-existence of maidens."

Fat chance of finding a maiden with you around, thought Merlin, but said nothing. From his bag he removed a large cloth, which he draped over his head. "I need total darkness for this spell," he explained. Safely concealed under its cover he hauled out a crystal ball and called up the shade of the great English puzzlist Henry Ernest Dudeney, which — under the threat of dire punishments from Merlin if it remained silent — revealed to the wizard Dudeney's secret "buttons-and-string" method.

Imagine a button placed on each square of the 4 × 4 board, and tie strings between any two buttons that are a knight's move apart. Rearrange the buttons, along with their connecting strings, to reveal the interconnections more clearly. Now to visit each button once, moving along the connecting strings; the problem is not changed by the way the strings are rearranged. More abstractly, we can say that the chessboard and its knight's-move interconnections are represented by a graph. The nodes of the graph correspond to the squares of the board, and an edge is drawn between two nodes when the corresponding squares are a knight's move apart. Then the knight's quest becomes a Hamiltonian circuit of the graph, that is, a path that travels along edges (strings) visiting each node (button) exactly once. Such circuits are named after the Irish mathematician William Rowan Hamilton. Hamiltonian circuits are not well understood and continue to be the subject of mathematical research.

Merlin showed Arthur and Laughalot the resulting diagram for the 4 × 4 knight's quest (Figure 45). Observe that there are four outer buttons, 1, 4, 13, 16; four middle buttons, 6, 7, 10, 11; and two *inner*

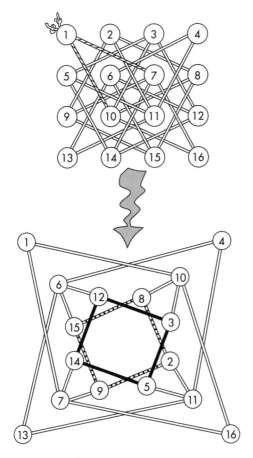

FIGURE **45**   Buttons-and-string proof that there is no 4 × 4 knight's quest.

squares—3, 5, 12, 14 and 2, 8, 9, 15. These inner squares are shown by a heavy line and a heavy dotted line, and will play an important role.

"Your majesty, I begin by arguing that a re-entrant quest is not possible; moreover, if there be a quest then it must begin and end on outer buttons."

"Why, Merlin?"

"My lord, let us suppose that there existeth a re-entrant quest. It must pass through button 13. But only buttons 6 and 11 are attached to 13, so the quest must contain the sequence 6–13–11 or 11–13–6. Similarly button 4 must be in the middle of a sequence 6–4–11 or 11–4–6. However, these parts of the supposed quest join up to form a closed loop 6–13–11–4 and back to 6. But a full quest visits each of the 16 squares, and cannot contain a closed loop of length 4. Thus a re-entrant quest is impossible. Either button 13 or 4 is at one end of the quest. Similarly either button 1 or 16 is at the other end."

The discussion was interrupted by Sir Garrulous, arriving with a dead basilisk over one arm and a disheveled damsel on the other. He introduced the damsel as the Lady Migraine. But Merlin, undaunted by such distractions, was still pursuing his proof.

"For definiteness, suppose the quest starteth at button 13 and endeth at 1. Then the quest starteth as, let us say, 13–11–4–6 and endeth as 7–16–10–1. There are other possible choices, but the argument that follows applies equally to them: the chosen paths are typical. Thus we must determine a subquest between two middle buttons 6 and 7 that passeth not through any of the buttons already used. These are the outer buttons 1, 4, 13, 16 and the other two middle buttons 10, 11. We thus require a subquest that runneth from button 6 to button 7 *and visiteth only the eight buttons on the two inner squares.*"

"Those buttons are quite *dreadful*," said the Lady Migraine. "The color doesn't match my eyes, Garry dear. I think we should—"

"However," Merlin plunged on, "there are no direct connections between one inner square and the other! Starting from 6 we must move to an inner square, let us say the solid one, by jumping from 6 to 12. But we can only get off that inner square by making an exit from 14 to 7. However, there is now no way to get on to the other (dotted) inner square, because button 7 is already connected to button 16! The same argument holdeth if we start from 6 and move to the dotted square: the only way to get on to the solid one is to pass through square 7, and that is not possible since we want the subquest to *end* at button 7.

"Thus, a 4 × 4 quest is impossible," he ended in triumph. There was a long silence, punctuated only by loud snores from Sir Garrulous. The moment was starting to become embarrassing when the king spoke. "Of course," he said airily. "I knew tha' before thou began thy speech." Twenty-twenty hindsight is ever the prerogative of royalty, and Merlin made no demur. He was used to it.

Laughalot was now hopping urgently from one leg to the other. "My liege, I really *must* convey to you some most important tidings—"

But Arthur had suddenly had an idea of his own. So unusual was the experience that he was quite overcome with excitement. "If a full ques' on a 4 × 4 ches'board is impossible," he said, "I wonder wha' the longes' possible ques' *is?*"

PROBLEM ❸

Can you answer Arthur's question? Sir Laughalot couldn't, so he had to keep quiet.

Knights may also quest on boards that are not square. The simplest such boards are rectangles. No complete quest is possible on a 2 × 3 rectangle, but a re-entrant quest is possible on a 3 × 4 rectangle. You can also use the buttons-and-string method to show that no quest is possible on the 3 × 5 or 3 × 6 rectangle.

PROBLEM ❹

Quests of the 3 × 7 rectangle are possible, but the choice of starting or finishing square is severely limited. How?

The knight may also roam in three dimensions rather than two. In his book *Amusements in Mathematics* Dudeney says: "Some few years ago I happened to read somewhere that Abnit Vandermonde, a clever mathematician who was born in 1736 and died in 1793 . . . had proposed the question of the quest of a knight over the six surfaces of a cube, each surface being a chessboard." Dudeney found a solution in which each face is quested in turn.

PROBLEM ❺

Can you do likewise?

Dudeney's memory, incidentally, was a little faulty. The name was Alexandre-Théophile Vandermonde, and he lived from 1735 to 1796.

Are similar quests possible on other sizes of cube? On a $2 \times 2 \times 2$ cube it is obviously impossible to quest each face in turn, but despite the lack of space, a re-entrant quest is possible. Before showing how to find one, we must specify more carefully how to move a knight on a surface. We consider a move to be legal if, *when the surface is flattened out in some way*, the pattern is the same as a standard knight's move. On a $2 \times 2 \times 2$ cube this produces a surprise. Figure 46 *A* shows a move that at first sight does not look like a knight's move, but which can be seen to be one if the surface is opened out flat, as in Figure 46 *B*.

Indeed, there is a re-entrant knight's quest on a $2 \times 2 \times 2$ cube (Figure 46 *C*), composed largely of these "surprise" moves. It has connections with an old puzzle: how can you slice a cube to produce a regular hexagon? There are four ways to do this, and each hexagon passes diagonally through six of the squares on a $2 \times 2 \times 2$ cube (Figure 46 *D*). Thus each hexagon defines a chain of six squares, linked by surprise knight's moves (Figure 46 *E*). It is not hard to find linking moves between these four surprise hexagons that join them all up in a single re-entrant quest (Figure 46 *F*).

So despite the lack of room on a $2 \times 2 \times 2$ cube, a knight's quest is possible. Surely, however, the $1 \times 1 \times 1$ cube is too cramped? Not a bit of it! Again there are surprise moves, which this time pass from a given square to any adjacent square. On a $1 \times 1 \times 1$ cube, the knight becomes a rook! A quest for this knight in rook's clothing is then so easy to devise that I will not draw it.

After all this had been explained, King Arthur finally allowed Sir Laughalot to convey his important news.

"Your majesty, I was attempting to tell thee that a horde of serpents whose gaze turneth living beings to stone hath invaded Romney Marsh. The peasants are fleeing but many will be caught."

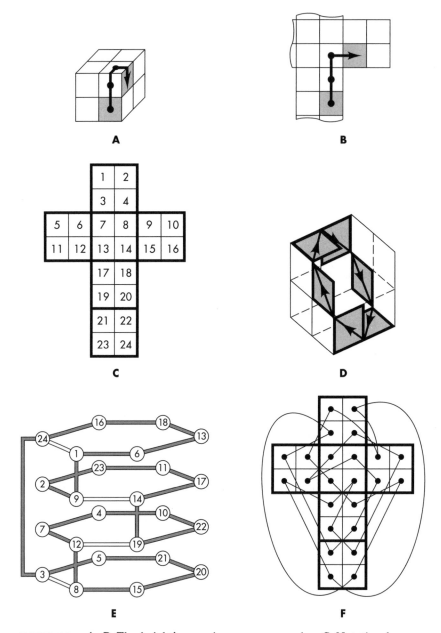

FIGURE **46** *A, B.* The knight's surprise move on a cube. *C.* Notation for squares. *D.* A surprise heaxagon. *E.* Building a knight's quest over a $2 \times 2 \times 2$ cube out of four surprise hexagons. *F.* The resulting solution. The cube has been flattened out for clarity.

The king thought about this for a moment. Then Laughalot said with resignation, "In any case, 'tis now far too late."

"Never mind," said the king. "There are plenty more peasants."

"I agree, your majesty. But what do we do with twenty thousand stone sheep?"

Arthur felt that something to distract his subjects' attention might be necessary while a troop of his knights quietly dropped the stone sheep into the sea, so he decreed that Merlin should arrange a great Ches' Tourney. And of course he did, and it worked, though some subjects did observe a mysterious rise in sea level. The ches' tourney was a great success, for Arthur had decreed this, too, and who was to gainsay the king? But afterward the Uther Pendragon Jousting Arena and Fishmarket had to be cleared, to make way for some new beasts of burden that Sir Capstan, returning from the crusades, was rumored to be bringing from far Arabia. King Arthur felt that the arena should be given a new name, to fit its new purpose, but could not think of one. He therefore proclaimed that a prize of half a gold crown would be given to the first man or woman who found an appropriate name.

Meanwhile, he was short one ches'board. But not for long.

Everybody has heard of Arthur's round table. But the King was so pleased with his invention (as he maintained) of the game of ches' that he had it replaced by a square one, divided up into an 8×8 checkered pattern. Merlin, fuming quietly at having his idea pinched, surreptitiously cast a spell on the table so that its opposite edges seemed to join together. He was of course unaware that he was thereby anticipating certain constructions in topology. Topologists form a variety of surfaces by bending pieces of the plane and gluing the edges. If the two ends of a rectangular strip of paper are glued together, the result is a cylinder. If they are twisted and glued, you get a Möbius band. If opposite edges of a rectangle are glued together, the result is a torus. Finally, if the opposite edges of a rectangle are glued together, but one pair is twisted, you get the curious Klein bottle—a surface with no edges but only one side! (Figure 47).

Mathematically, it isn't actually necessary to perform the bending: you just have to imagine that the appropriate edges are next to each other when working out how to move across an edge. In this way the rectangle remains flat, in the sense that you can pave it with perfectly regular squares, even though the usual ways to draw surfaces such as a

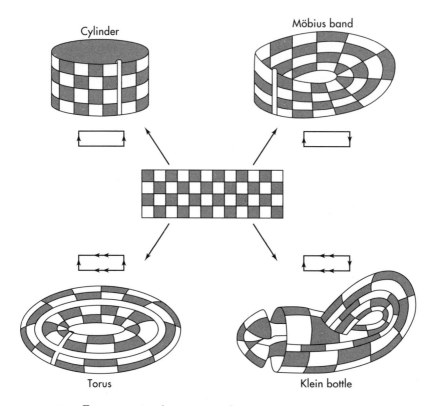

Cylinder

Möbius band

Torus

Klein bottle

FIGURE **47** Four ways to glue a rectangle.

torus appear curved. A rectangle with its edges identified forms a flat torus. And because the torus is flat we can mark out ches'boards on the rectangle, and think about knight's quests on cylinders, Möbius bands, tori, and Klein bottles.

Arthur was a little perturbed by the geometry of his new table, but Merlin told him (truthfully but misleadingly) that an evil sorcerer had placed it under an enchantment. Which they would have to live with until he, Merlin, could come up with an effective counterspell. Which he would work on, provided the king was willing to fund a research project to the tune of five thousand gold crowns. In the meantime, he went on, the company would be able to improve its intellectual qualities by contemplating the mysteries of cylinders, Möbius bands, Klein bottles, and tori.

Sir Garrulous immediately ran into trouble thinking about Klein bottles. "I can't see how to make it unless you let it pass through itself," he told Merlin one day. "Which seems to be cheating a bit," he added apologetically.

Merlin agreed that the Klein bottle cannot be constructed in three-dimensional space without passing through itself. "But," he declaimed, "that does not prevent us considering knight's quests on a Klein bottle ches'board!"

"Why no'?" asked Arthur.

"We shall see," said Merlin mysteriously. "But let us begin with cylinders. To visualize the quest I shall place 'ghost' copies of the basic rectangle at each end and pretend that corresponding cells are the same as those in the original rectangle. The knight may then move off the edge on to a ghost, provided it is immediately replaced at the corresponding position of the original rectangle. Quests on a $2 \times n$ cylinder or Möbius band are possible only when n is odd. Quests on a $3 \times n$ and $5 \times n$ cylinder are always possible, using a simple repetitive pattern. Moreover, several such cylinders can be joined edge to edge, and the quests combined by breaking them at suitable places and rejoining them [Figure 48]. The height of such a cylinder can be any number of the form $3a + 5b$. Only the numbers 1, 2, 4, 7 are not of this form, so I have proved that a knight's quest is possible on an $m \times n$ cylinder for all m except perhaps 1, 2, 4, 7. Quests on a $1 \times n$ cylinder are impossible, and I have already discussed $2 \times n$ cylinders."

"And what of the other two?" asked Laughalot. But Merlin had a ready reply, a favorite among hard-pressed mathematicians who have forgotten what the answer to a problem is. "As an exercise I leave $4 \times n$ and $7 \times n$ cylinders up to you," he said grandly. "Together with similar questions for Möbius bands."

"Personally," said Laughalot, "I'd rather attack a nest of dragons with a toothpick." But maybe the reader is made of sterner stuff.

Merlin was by now under full steam and impossible to stop. "Quests of the torus," he said, "can also be visualized by placing ghost rectangles beside the original. Although the 4×4 square has no knight's quest, there is a re-entrant question on the 4×4 torus. There is a quest, but not a re-entrant one, on the 4×4 cylinder. To see why, I again use the

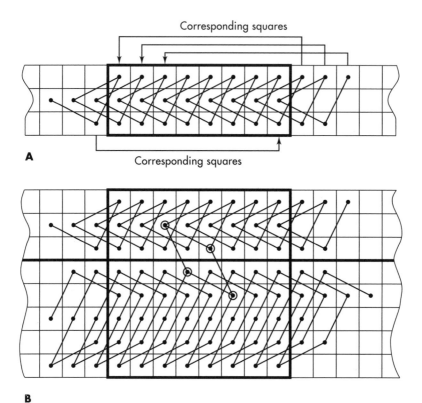

FIGURE **48** *A.* Quests on a cylinder can be visualised by adding ghosts on the ends and identifying corresponding squares. *B.* Quests on several cylinders can be joined together by changing the links in a suitable parallelogram.

buttons-and-string method. For a 4×4 torus the resulting graph is exactly the same as the vertices and edges of a hypercube — the 4-dimensional analogue of a cube [Figure 49 *A*]. The knight's quest on a 4×4 torus is really a quest in four dimensions! From the graph it is easy to find a Hamiltonian circuit yielding the re-entrant quest illustrated." He stopped for breath, but before anyone could get a word in he was off again. "For a 4×4 cylinder, the graph is similar, but several edges are deleted. The top and bottom faces of the hypercube remain intact, but only four vertical edges remain [Figure 49 *B*]. I have a clever

1	2	3	4	1	2	3	4	1	2	3	4
5	6	7	8	5	6	7	8	5	6	7	8
9	10	11	12	9	10	11	12	9	10	11	12
13	14	15	16	13	14	15	16	13	14	15	16
1	2	3	4	1	2	3	4	1	2	3	4
5	6	7	8	5	6	7	8	5	6	7	8
9	10	11	12	9	10	11	12	9	10	11	12
13	14	15	16	13	14	15	16	13	14	15	16
1	2	3	4	1	2	3	4	1	2	3	4
5	6	7	8	5	6	7	8	5	6	7	8
9	10	11	12	9	10	11	12	9	10	11	12
13	14	15	16	13	14	15	16	13	14	15	16

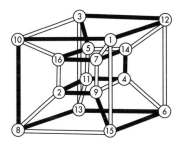

A

1	2	3	4	1	2	3	4	1	2	3	4
5	6	7	8	5	6	7	8	5	6	7	8
9	10	11	12	9	10	11	12	9	10	11	12
13	14	15	16	13	14	15	16	13	14	15	16

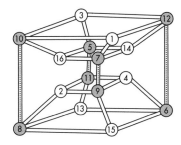

B

FIGURE **49**   *A.* The 4 × 4 torus surrounded by ghosts; the corresponding hypercubic graph; and a Hamiltonian circuit (heavy lines) yielding a knight's quest. *B.* The 4 × 4 cylinder plus ghosts, and its graph, shaded to prove that no re-entrant knight's quest exists.

proof that no re-entrant quest is possible—let me tell you about it." Lady Migraine groaned, but Merlin pretended not to hear. "I shade the buttons corresponding to squares 5, 7, 10, 12 and 6, 8, 9, 11, as shown. Observe that any move starting at a white button must end on a shaded button. Most moves starting on shaded buttons end on white ones; but there are four exceptions. These are the vertical links 5–11, 7–9, 10–8, and 12–6. If a re-entrant quest exists, then it must pass through every button. There are eight white and eight shaded buttons, and each white button must be followed by a shaded one. Hence the colors must alternate. However, to pass from the top face of the hypercube to the bottom one we must use one of the four exceptional links, where the color does not change. Thus it is impossible to make the colors alternate, and therefore no re-entrant quest exists. However, non-re-entrant quests *do* exist; and the same method of shading shows that they must start and end on white buttons, one in the top face and one in the lower face."

To the accompaniment of Sir Garrulous's sonorous snores, renowned throughout six neighboring kingdoms, Merlin moved on to the corresponding questions for the Klein bottle, for which alternate columns of ghosts must be turned upside down. He explained at length why many cases pose no new difficulty. For example, if an ordinary knight's quest is possible on an $m \times n$ rectangle, then the same quest works when edges are glued, so (for example) the 6×6 knight's quest also solves the problem for a 6×6 cylinder, Möbius band, torus, and Klein bottle. For similar reasons, if a quest is possible on an $m \times n$ rectangle arranged in the form of a cylinder, then it must also be possible on a torus and a Klein bottle of those dimensions.

"Now in four-dimensional space—"

Sir Laughalot gave a wild yell and ran from the room.

"Where art thou off to?" the king shouted toward the retreating back. Laughalot stopped in his tracks and turned.

"I've just remembered an urgent appointment," he said. "I've got to go and find—er—the—um . . ."

"The—the Holy Grail!" cried Sir Garrulous, who had been awoken by Laughalot's yell. "I've got to find that, too! What an *amazing* coincidence! Wait for me!" Together they clanked away across the stone-flagged courtyard.

"Well," sneered the Lady Migraine. "I always thought that Sir Garrulous wasn't to be trusted! Off to see another damsel in distress, I bet! Why, that rotten no-good—"

It was clear that Lady Migraine's diatribe was only beginning. Arthur looked helplessly at Merlin, who had clasped his hands over his ears. He was just reaching the point of desperation, and wondering whether to risk the spell known as Furnwold's Erratic Lip-weld, when a page sounded a trumpet and a tall figure clad in black armor strode into the court and bowed low before the king.

"Your majesty, I seek a boon."

"Arise, Sir Perimason. Wha' boon?"

"Thy proclamation. The quest for a name for the arena to house the beasts of burden being brought from the Arabian crusades by Sir Capstan."

"Ah," said the king, who had forgotten all about it. "Well, ou' with i', man! What' name has' thou devised?"

Sir Perimason cleared his throat. "Camel Lot," he said.

ANSWERS

1. Any re-entrant quest must visit equal numbers of black and white squares. On a 5 × 5 or 7 × 7 board—or indeed any board with an *odd* number of squares altogether—a re-entrant quest is therefore impossible.

2. No quest is possible on the diamond-shaped board of Figure 44. To see why, color the squares alternately black and white, like a standard chessboard. Observe that the knight changes the color of its square on each move. Thus, either the number of black and white squares is the same, or there is one more black than white, or one less. But the diamond-shaped board under discussion has 32 black and 37 white squares. Thus, the longest quest that might be possible visits 65 squares, 32 black and 33 white, and starts and ends on white squares. Does such a quest exist? (I'm not going to give an answer this time.)

3. The longest quest on a 4 × 4 board visits 15 squares, for example, 1−7−16−10−3−5−14−12−6−4−11−2−8−15−9.

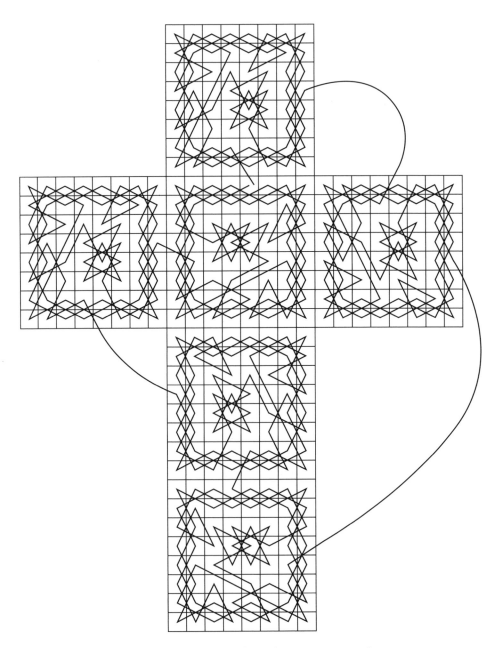

FIGURE **50**   Dudeney's tour on the surface of an $8 \times 8 \times 8$ cube.

4. On a 3 × 7 rectangle, the starting or finishing square must be one space diagonally inward from a corner.

5. A quest over the surface of a cube is shown in Figure 50. There are many other solutions.

FURTHER READING

Ball, W. W. Rouse. *Mathematical Recreations and Essays*. London: Macmillan, 1959.

Dudeney, H. E. *Amusements in Mathematics*. New York: Dover, 1958.

Gardner, Martin. *Mathematical Magic Show*. New York: Alfred A. Knopf, 1977; Harmondsworth, England: Penguin Books, 1985.

Kraitchik, Maurice. *Mathematical Recreations*. London: Allen and Unwin, 1960.

8

A Vine Math You've Got Me Into

There I was in a ramshackle bus backfiring its way along the road to Frascati to join my friends Elena and Enrico Macaroni at Enrico's brother's vineyard. After Enrico had introduced me to his brother Alberto, it turned out that they had an ulterior motive. I had been invited for my mathematical expertise. We playfully negotiated a "consultancy fee." It came by the crate; I'm very partial to Frascati.

"So, Alberto—what's the problem?"

"Teroldego," said Alberto.

"Sorry, but I don't speak—"

"Nebbiolo, Trebiano, Malvasia, Moscato, Aleatico, Sangiovese," he added.

"What?"

"Grapes. They are grapes. I have seven varieties of grape to test, to see which gives the best wine. I wish to plant them in plots on the hillside. Unfortunately the hill is narrow and I can plant only three varieties of grape on each plot of land. However, I want to minimize the effects of different soils and different exposure to the sun."

"Very sensible," I said knowingly. "Good experimental design is very important to eliminate error."

"Quite. I have drawn up some requirements which I believe achieve these aims." With a flourish he produced a sheet of paper, on which was written:

> *Seven* varieties of grapes are to be arranged in plots.
> Each plot contains exactly *three* different varieties.
> The following conditions must hold:
>   *A.* Any two plots have exactly one variety in common.
>   *B.* Any two varieties lie in exactly one common plot.

"Very clearly expressed," I told him. "What's the problem?"

"I am unable to find a suitable arrangement that satisfies the two conditions," he said.

I didn't say anything, but it did seem to me that he hadn't tried very hard. *Before reading on, see if you can find a suitable arrangement.*

"Never fear," I said, "Stewart is here. Your difficulties are at an end. I know a solution."

"Excellent!" said Alberto.

"Magnifico!" cried Elena. "For once, my friend, you are making yourself useful, instead of running off at the mouth about those stupid protected drains."

"*Projective planes**, Elena."

"That's what I said. I cannot *abide* your protected drains. Alberto, thank heaven that your problem is not about geometry! This crazy man will stuff your mind with parallels that meet at infinity, circles that are straight lines, and surfaces that have only one side!" She turned her head my way. "Now, Ian, tell us your answer."

"I hate to say this—" I said.

"Oh, no."

---

*See *Game, Set, and Math*, Chapter 11.

"It—um—involves projective planes," I said. "Finite projective planes. You see, it *is* a geometric problem, but one involving finitely many points."

"This is no improvement. He has gone batty again," said Elena sadly, addressing thin air. I decided to maintain my dignity and simply continue.

"Your three conditions are very similar to what happens in projective geometry," I said. "Suppose I replace 'plot' by 'line' and 'variety of grapes' by 'point.' Then your conditions become:

Seven points are to be arranged in lines.
Each line contains exactly *three* different points.
The following conditions must hold:
  A. Any two lines have exactly one point in common.
  B. Any two points lie in exactly one common line.

See? It's geometry."

"Hmmm. But grapes, even though they are small and round, are not points. And plots of land, even if long and narrow, are not lines."

"True. But irrelevant. We're talking about abstract properties of *arrangements* of objects, not the objects themselves. Logically, it makes no difference what *names* we call them. As the great German mathematician David Hilbert said, the logical structure of geometry should make equal sense if the words 'point,' 'line,' 'plane' are replaced by 'mug,' 'chair,' 'table.' Names don't matter."

"'The penguins at the chimney of an isosceles concertina are equal . . .' Yes, it does have an aura of freshness—"

"*Logically* speaking, names don't matter. But *psycho*logically, geometric names remind us of analogies with genuine geometric objects. Here we get a form of geometry in which there are no parallel lines. *Any* two lines have a point in common. That condition fails in *Euclidean* geometry, but it holds in *projective* geometry.

"Which began in Italy," I continued hurriedly, seeing the skeptical look in Alberto's eye. "It dates from the discovery of the laws of perspective by Brunelleschi. And *Alberti*," I added.

The look faded. "But of course," said Alberto. "Euclid, he was Greek, no? We need *Italian* geometry for an Italian problem." Italian culture had partially convinced him, and his near-namesake clinched it.

"For god's sake," Elena stage-whispered, "*Don't* get him on to Steiner's Roman surface or he'll start that nonsense† all over again!"

"Have no fear, Elena my dear," I said. "Finite projective planes are a branch of combinatorics, not topology. Alberto, let me draw the answer to your problem." And I sketched a diagram (Figure 51).

"Very interesting," said Alberto. "But one line is *bent*," Enrico protested. "And I don't see—"

"Sorry, let me interpret the diagram in terms of the original problem. If I list the triples of 'points' on each 'line' I get this:

$$\begin{array}{ccc} 1 & 2 & 6 \\ 1 & 3 & 5 \\ 1 & 4 & 7 \\ 2 & 3 & 4 \\ 2 & 5 & 7 \\ 3 & 6 & 7 \\ 4 & 5 & 6 \end{array}$$

The numbers 1–7 are the varieties of grape, and the list represents seven different plots on the hillside, each containing three varieties. If you check you'll find that both your conditions hold.

"'*A.* Any two plots have exactly one variety in common.' For

---

†See *Game, Set, and Math* Chapter 11.

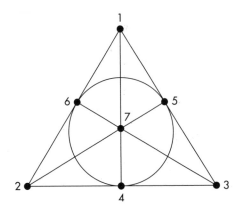

FIGURE 51 A finite projective plane with three points on each of seven lines.

LINES BETWEEN PAIRS OF POINTS

first point		1	2	3	4	5	6
second point	2	126					
	3	135	234				
	4	147	234	234			
	5	135	257	135	456		
	6	126	126	367	456	456	
	7	147	257	367	147	257	367

example, the first two plots 126 and 135 have just 1 in common. Plots 234 and 456 have just 4 in common, and so on.

"'1*B*. Any two varieties lie in exactly one common plot.' For instance, varieties 1 and 5 lie in 135, and in none of the other plots. And varieties 3 and 6 lie in 367 only.

"In fact, I can draw up two tables that check all possible combinations" (see the tables "Lines Between Pairs of Points" and "Points Where Lines Meet").

"But the *curved* line—"

"It has to be drawn in a curve because the 'points' and 'lines' aren't *real* points and lines, they're varieties of grapes and plots of land! Look at the list! It works! Do you *care* that one 'line' is bent?"

POINTS WHERE LINES MEET

first line		126	135	147	234	257	367
second line	135	1					
	147	1	1				
	234	2	3	4			
	257	2	5	7	2		
	367	6	3	7	3	7	
	456	6	5	4	4	5	6

"I guess not."

"But the diagram *is* useful," I said. "A lot easier to remember than a list. And you can *see* that it satisfies all three conditions. You don't really have to check the tables.

"Right. Great. I'll tell Pigro to get planting right away. Oh, and do you mind if I telephone cousin Vittorio? He has a problem similar to mine, maybe you could solve that, too."

"The more the merrier," I said expansively. "Provided the fee is of comparable quantity and quality!"

"Excellent! And while we wait for Vittorio to arrive, let me serve you a glass of local wine, from Montefiascone, which has the strangest name of all wines!" He disappeared and came back with a bottle. I looked at the label.

"What sort of a name is *Est Est Est?*" I asked.

"The name, I regret, is the most memorable characteristic of the wine," said Alberto. "Though it is palatable on a hot day. Many years ago a clergyman by the name of Bishop Fugger was traveling to Rome, and he sent a servant ahead to find out which inns had the best facilities (by which he meant wine, for all beds were lumpy and all food inedible). The servant would chalk the Latin word *est*—meaning 'is'—on the door of an acceptable inn, and *non est*—'is not'—on the rest. One evening the servant arrived, dirty and travel-sore, in Montefiascone. It looked a real dump. But when he ordered wine, they accidentally gave him a good bottle. He was so impressed that he wrote *Est! Est! Est!* on the door, and—"

"Come on, now."

"No, really, every word is true—well, most, anyway. The servant invented the name, and he also invented the three-star hotel rating!" A noise made him turn in his seat. "Aha! Here is cousin Vittorio!"

Much embracing and back-slapping later, we got down to business. It turned out that Vittorio, too, had some grapes to test.

"But I have thirteen varieties, rather than seven. And my plots are larger: each can hold four different varieties. But I still want Alberto's two rules to hold."

"Terrific," I said, thinking rapidly. *Thirteen points, in lines of four, each pair of lines meeting in a unique point and each pair of points lying on a unique line. That's another projective plane!*

I told them I could solve the problem. "But," I said, determined to extract maximum advantage, "I'll only tell you if you let me tell it *right*. I want to explain where the answer comes from, as well as what it is."

"Why?"

"First, because answers without reasons are magic, not mathematics. Second, because knowing why that answer occurs can help you solve similar problems later. Agreed?"

"Agreed," said Vittorio. Elena sighed and Alberto called for another dozen bottles of *Est Est Est.*

I settled down for a long afternoon. "The starting point is the connection‡ between the usual Euclidean plane and the projective plane. To get the projective plane you take the Euclidean plane and add an extra 'line at infinity' which has one point for each *direction* in the Euclidean plane. If a bunch of parallel lines all point in the same direction, they are deemed to pass through—and hence meet at—the corresponding point on the line at infinity [Figure 52]. OK so far, Vittorio?"

---

‡See *Game, Set, and Math,* Chapter 11.

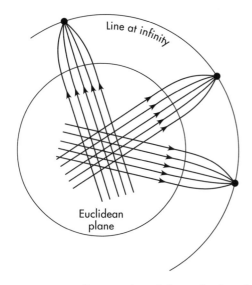

FIGURE **52**   Construction of the projective plane by adding "points at infinity" to the Euclidean plane, corresponding to sets of parallel lines.

"I think so."

"Good. Now, we can specify points on the Euclidean plane using two *coordinates x* and *y*, which are real numbers. In fact, you get a coordinate grid labeled by the two numbers *x* and *y*. To get a *finite* projective plane we start by getting a finite Euclidean plane. Actually, the technical term is affine plane. We do that by selecting coordinates from something different from the real numbers. For example we might use just two coordinate values, 0 and 1. That gives four points, namely (0,0), (0,1), (1,0), and (1,1). We can draw them as a square [Figure 53A].

"Now we have to decide what the 'straight lines' in the diagram should be. In this case it's easy: the sides of the square, and its diagonals [Figure 53B].

"Now we mimic the construction of the real projective plane, by adding a 'line at infinity.' The top and bottom of the square are 'parallel,' in the sense that they don't meet. So we add a point 'at infinity' on to each. Similarly for the left and right sides. The diagonals form a third pair of 'parallel' lines."

"But they meet," Enrico protested.

"Not on the grid," I said. In *this* plane only the numbers 0 and 1 exist. The diagonals meet at $(\frac{1}{2}, \frac{1}{2})$ and that doesn't count. Since they don't meet *on the grid*, I consider them to be parallel."

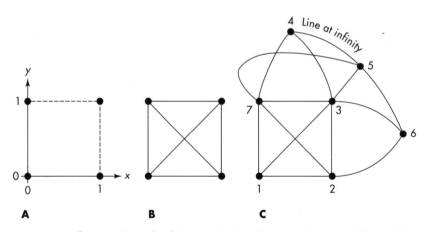

FIGURE 53 Construction of a finite projective plane, analogous to Figure 52. *A*. Initial grid is a finite analogue of the Euclidean plane, using only 0 and 1 for coordinates. *B*. Lines in the grid. *C*. Adding a line at infinity, with one point for each set of "parallels" in the grid.

"Well . . ."

"I'll make it seem a bit less arbitrary in a moment. Anyway, the three sets of parallels provide three extra points, forming a new line — 'at infinity' — and we get a system with seven points and seven lines [Figure 53*C*], which we've seen before, but drawn in a different form. If you compare the numbers on the points you'll see you get the same list of lines that we first got" (Figure 51).

"Wonderful!" yelled Alberto. "I think," he added.

"Really, to see why we think of the diagonals as 'parallel,' you have to imagine that the opposite sides of the square 'wrap around,'" I said.

"Yes, but . . ."

"Why wrap around? Well, that's a long story: homogeneous coordinates and finite fields —"

"Is a vineyard, not a field!"

"No need to get excited, Vittorio! 'Field' is a technical term meaning a system in which you can add, subtract, multiply, and divide, while keeping all the usual laws of algebra. The coordinates in the Euclidean plane are real numbers. You can add and multiply the coordinates to get other real numbers. Moreover, in coordinate geometry the condition for two lines to be parallel involves the algebraic structure, so there's a link between geometry and algebra.

"If you want to establish a similar link for a system of coordinates that contains only the numbers 0 and 1, you must insist that the sums and products are *also* confined to the values 0 and 1. There's no problem with

$$0 \times 0 = 0 \quad 0 \times 1 = 1 \quad 1 \times 1 = 1$$

or

$$0 + 0 = 0 \quad 0 + 1 = 1$$

but you can't have $1 + 1 = 2$ because that goes outside the set 0 and 1. So instead you have to agree that

$$1 + 1 = 0$$

"Why?"

"The only other possibility would be $1 + 1 = 1$, but then you could subtract 1 to get $1 = 0$, which is ridiculous."

"Oh, of course. Ridiculous. But $2 = 0$ is perfectly sensible. Why did I not see—"

"This system doesn't *have* a number 2."

Vittorio looked unhappy. "Now you are saying that $1 + 1 = 2$ is not one of the 'usual laws of algebra'?"

"It's usual for real numbers but not for other systems. If you like, you can think of 0 as meaning 'even' and 1 as 'odd.' The odd + odd = even, you see, so $1 + 1 = 0$ is correct."

"Mad," said Elena, tapping her head with a finger.

"No, I'm talking about the algebra of finite fields," I said. "But this is neither the time nor place to talk about those. All I really want you to believe is that the way to draw straight lines on a finite grid is to make them 'wrap around.'

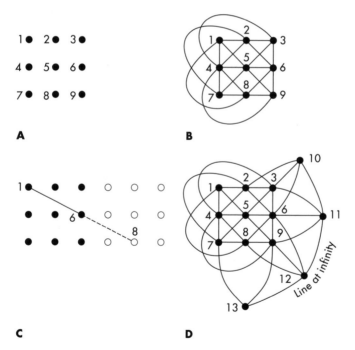

A   **B**

C   **D**

FIGURE 54 Construction of a finite projective plane with 13 points. *A.* Start with a 3×3 grid. *B.* The four sets of parallels. *C.* Why the line from 1 to 6 is not new: wrapped around it extends to include point 8, but 186 is already included. *D.* The finished article!

"The 2 × 2 grid is so simple it's misleading. Everything becomes much clearer if we start with a system in which the horizontal coordinates are 0, 1, and 2. Then we get a 3 × 3 grid of nine points [Figure 54A]. This corresponds to a finite field with the three elements, 0, 1, and 2, in which addition and multiplication are defined 'modulo 3.' That is, you throw away multiples of 3. So $1 + 2 = 0$, $2 + 2 = 1$, $2 \times 2 = 1$, and so on.

"This number system 'wraps around' from 3 to 0. Similarly, the lines in the grid wrap around. They come in sets of three 'parallel' lines [Figure 54B]. The obvious ones are the horizontal lines 123, 456, 789 and the vertical ones 147, 258, 369. But there are two more sets — the ones which wrap around — 'broken diagonals' slanting one way, 159, 267, 348, and 'broken diagonals' slanting the other way, namely 753, 429, and 186."

"What about lines with different slopes? For instance, the line from 1 to 6? That doesn't point horizontally or vertically or diagonally!"

Vittorio was getting the hang of it, I could tell. "If you continue that line and let it wrap around you'll find it passes through point 8. It's just the diagonal 186 in disguise" (Figure 54C).

"Now all you have to do is add a 'line at infinity' with four points, numbered 10, 11, 12, 13, one for each set of parallels. Then you get a 13-point projective plane [Figure 54D]. And you can read off a list of 13 plots, each with four varieties of grapes in it, that solves Vittorio's problem, like this:

1	2	3	11
4	5	6	11
7	8	9	11
1	4	7	13
2	5	8	13
3	6	9	13
1	5	9	12
2	6	7	12
3	4	8	12
7	5	3	10
4	2	9	10
1	8	6	10
10	11	12	13	(line at infinity)

I've grouped them together according to the four directions determined by the sets of three 'parallels,' and put the line at infinity at the end"

"That's amazing!" said Vittorio. "Let me try! I'll start with a 4 × 4 grid!"

"Whoops, no, you—"

"I can do it! Don't interrupt, I'll soon sort it out. . ."

Two hours later, the sun was beginning to set, and Vittorio's eyes had a glazed look.

"It doesn't work," he said finally. "I either get too many points on lines or not enough lines through points. Look, if I make lines wrap around then I can find distinct lines that meet in two points, not one!" (Figure 55).

"I was trying to tell you that," I said. "If you use the same idea on a 4 × 4 grid it doesn't work properly. You can't get the sets of 'parallels' to behave sensibly. That, it turns out, is because 4 isn't prime. You have to use a square grid whose size is a prime number. Then it always works. So you can start with a 5 × 5 grid, getting six sets each containing five 'parallel' lines, and end up with a projective plane containing 31 points arranged in lines of 6."

PROBLEM ❶

Vittorio rapidly drew such a projective plane. Can you?

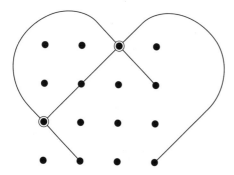

FIGURE 55   The 4 × 4 grid hits a snag. Here two lines meet in more than one point.

"Where did those numbers come from?" Alberto asked.
"It's a general rule. We have

$$7 = 2^2 + 2 + 1$$
$$13 = 3^2 + 3 + 1$$
$$31 = 5^2 + 5 + 1$$

That's not a coincidence. If you have a projective plane of *order n*, meaning that there are $n + 1$ points on each line, then the number of points altogether must be $n^2 + n + 1$ [see the box below entitled "Numerology of Projective Planes"].

"I said that the grid method works when the size of the square— that is, the order—is prime. In fact, the method can be generalized to produce projective planes for any order that is a prime power. That's because there exist finite fields of all prime power sizes. So there exist

NUMEROLOGY OF PROJECTIVE PLANES

Suppose each line in a projective plane contains $n + 1$ points. Take any line L and any point P not on that line. Join P to each point of L (Figure 56, black dots). The resulting $n + 1$ lines are distinct; each contains P, some point of L, and $n - 1$ other points. Thus we have obtained a total of

$n + 1$	points on L
$+ 1$	point P
$+(n + 1)(n - 1)$	points on the lines from L to P

which adds up to $n^2 + n + 1$.

We now have to show that there are no other points in the projective plane. But if Q is any point, then there is a line M through P and Q. Furthermore, M and L must meet in some point R. But then Q lies on the line joining P to R, and R is one of the points on L. (Figure 56, white dots).

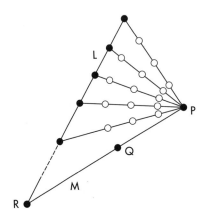

FIGURE 56   Counting points in a projective plane of order *n*.

projective planes of orders

$$2\ 3\ 4\ 5\ 7\ 8\ 9\ 11\ 13\ 16\ 17\ 19\ .\ .\ .$$

and so on. Even though you can't get an order 4 plane from a 4 × 4 grid, you *can* get one by starting with a slightly different arrangement of 16 points."

"What about the other orders?" asked Elena. "Six, for instance."

"Even *you're* starting to get interested now, aren't you?"

"Against my better judgment. It's much neater than those squirmy boy surfers you insisted on ramming down our —"

"*Boy's surface*, Elena. Beauty is in the eye of the beholder. But finite projective planes do have a certain charm. Anyway, you asked about the others, orders

$$6\ 10\ 12\ 14\ 15\ 20\ .\ .\ .$$

Those are *very* interesting.

"Around 1900 people knew that there was no order 6 projective plane. That is, you can't arrange 43 points in lines of 7 while obeying the conditions that each pair of lines meets in a unique point and each pair of points lies on a unique line. If Vittorio had had 43 varieties of grape to test, in plots that could hold seven varieties, he'd have been out of luck. In 1949 R. H. Bruck and H. J. Ryser proved the *only* known general result about the non-existence of finite projective planes. If the order *n* is of the form $4k + 1$ or $4k + 2$ and n is not the sum of two perfect squares, then no projective plane of order n exists. For example, 14 is $4 \times 3 + 2$ and is not a sum of two squares, so that rules out 14. The Bruck-Ryser theorem also rules out 6, 21, 22, and infinitely many other values."

"Remarkable."

"But that still leaves 10, 12, 15, 18, 20, and so on unsolved. For instance, when the order is 10, we have 111 points, to be arranged in lines of 11 apiece —"

"That is *such* a coincidence! My uncle Giorgio, he wants to test 111 varieties of grapes, and his plots can hold 11 each! He's a very rich man, my uncle Giorgio, and he has many big vineyards. Maybe you —"

"Wait! Hold your horses, Vittorio! I'm sorry to tell you that uncle Giorgio is in big trouble."

"You mean?"

"There is no order 10 projective plane."

"Who says?"

"A group of four mathematicians at Concordia University in Montreal — Clement Lam, Larry Thiel, Stanley Swiercz, and John MacKay. They used a supercomputer to do it, and it took them nine years. They say it took about a hundred times the computing power of another famous computer proof, Kenneth Appel and Wolfgang Haken's proof of the four-color theorem. You may have heard of it: you can color any map with four colors so that adjacent regions have different colors. That took a thousand hours of computing, so the Lam-Thiel-Swiercz-MacKay proof would have taken a hundred thousand hours on the same machine! Of course, they used a faster one —"

"These people, they better watch out. Uncle Giorgio, he will be angry."

"I'm sorry, but that's the way it goes. The order 12 case is open, by the way, but a computer attack along the same lines would take ten thousand million times as long."

"Uncle Giorgio, he will be *very* angry," Vittorio continued doggedly. "Alberto, he has a projective plane, and I, Vittorio, have a projective plane, but my extremely rich uncle Giorgio, he does not have a —"

"Oh, stuff your rich uncle Giorgio!" I yelled.

"I do not think you would want him to hear you say that," said Vittorio quietly, shaking his head.

"Why not?"

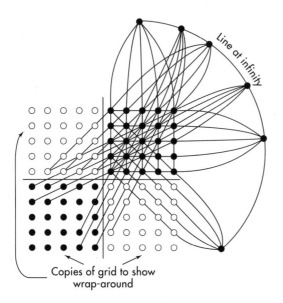

FIGURE **57** Projective plane formed from 5 × 5 grid. (It can be drawn in other apparently different ways: only the set of points on a given line counts, not the shape of the curve that links them.)

"Uncle Giorgio, he has—shall we say—contacts. Influential contacts."

"I'm not scared of politicians!"

"Not politicians," said Vittorio. "Not so—official. You see, uncle Giorgio, he lives in Palermo."

The light dawned. No, I certainly didn't want to offend the Mathemafia.

ANSWER

1. A projective plane of order 5, based on the 5 × 5 grid, is shown in Figure 57.

FURTHER READING

Batten, Lynn Margaret. *Combinatorics of Finite Geometries.* Cambridge and New York: Cambridge University Press, 1986.

Bruck, R. H., and H. J. Ryser. The nonexistence of certain finite projective planes. *Canadian Journal of Mathematics* 1 (1949): 88–93.

Cipra, Barry. Computer search solves an old math problem. *Science* 242 (16 December 1988): 1507–1508.

Marshall, Hall, Jr. *Combinatorial Theory*, 2d ed. New York: John Wiley & Sons, 1986.

●

Maxdoch Murwell,

●

Market Manipulator

●

●

"You can't beat market forces," said Rodney.

"I have no intention of trying to *beat* market forces," I said. "You can't *beat* thunderstorms either, but that doesn't mean you shouldn't put a roof on your house."

He ignored me and gestured expansively at our surroundings. "*That's* what market forces can achieve! The best possible outcome for everyone!" And I have to admit he had a point. The Mediterranean island of Taxhaven, until recently privately owned, had been opened up to tourists. The beach at Costa Narmanaleg is one kilometer long and straight as a die. Golden sand slopes down to an azure sea. I don't normally frequent such parts, being not so much a yuppie as a foswat (forty, overweight, sinking without a trace), but I was attending a conference on non-linear optimization at the Glitz Hotel. While walking along the sands I had bumped into an old acquaintance, Rodney Casshe-

Riche, and his latest girlfriend, Fiona. We had found a quiet spot among the rock pools near the western end. It was idyllic. Also hot.

There was an ice-cream vendor a few hundred meters away, the red and yellow sunshade on his little white refrigerated cart beckoning like a beacon.

"I'll just pop along and get us all some ice cream," said Rodney.

"Good idea," I said. "Make mine chocolate blue chip." Rodney had made a fortune speculating in junk bonds before they became illegal and took early retirement at the age of thirty-two.

"Hmmph," he snorted. "What about you, Fiona?"

"Nothing for me," she said, making a face. "I'm on a diet."

"I didn't notice that last night when we had the lobster and champagne," he said.

"It's a seafood diet."

Rodney and I looked at each other. Simultaneously we yelled, "When you see food, you eat it!" The old jokes are the best, I always say. It's true. That I always say it, I mean.

Rodney ambled over to the ice-cream man, who promptly took down his sunshade and closed the lid on his refrigerator. I wasn't surprised; Rodney has that effect on a lot of people. But this time it wasn't Rodney's fault, as became clear when he returned, a downcast look on his finely chiseled and normally supercilious features. "Temporarily closed," he said. "On account of the competition." The mayor had deregulated the town's economic system, and the whole island was a hotbed of private enterprise. It turned out that a second ice-cream salesman had set up shop at the other end of the beach. "Captured half his market at a stroke," said Rodney. "So he's relocating his premises, nearer to the middle, to increase his market share."

"Market forces," I said, "have just deprived us of our ice cream."

"Yes, but the overall service has improved," said Rodney. "Now there are two ice-cream men, so they serve the whole beach more efficiently because people don't have to walk so far to the nearest one. I bet the extra sales more than cover the cost of a second cart." The ice-cream man had now opened up for business again, about a quarter of the way along the beach and was surrounded by hordes of bronzed and screaming children. Maybe Rodney was right.

"Never mind, Ian," said Fiona. "There's some champagne in the ice bucket."

"Right now I'd settle for the ice. How about you, Rodney?"

"Stupid thing!" said Rodney. "No, not you, Fiona," he added quickly. "The carphone's on the blink!" It was portable, and he had detached it from the Lamborghini and brought it down to the beach. He gave a despairing cry and flung it into a rock pool. "Drat the confounded object! Oh, why did it have to go wrong *now*!"

"What's the matter?" asked Fiona.

"The takeover of Dross Bros.," he said. "I've been waiting for some inside information before deciding whether to buy or sell my shares, and now the phone's gone dead!"

"Then find another phone, dear."

He stomped off. Ten minutes later he stomped back. "Link on the blink," he said. "Satellite's transmitting but not receiving. I can phone out from anywhere on the island, but messages can't get in! I can buy or sell, but I can't find out what's happening!"

"Shame," I said. "The world's in a sad state when you can't get inside information about a takeover."

"Look, Ian," he said, "this is serious. I'm up against Maxdoch Murwell and if I jump the wrong way, I'll lose a fortune! I can't just sit and do nothing, he'll wipe me out! I need to make a decision and I haven't got the information I need!"

"Maybe I can help," I said.

"You? What do you know about the stock market?"

"Nothing."

"Ha!"

"But," I added, "I know a lot about decision-making." Rodney looked unimpressed, but I could tell he was hooked. The drowning man, clutching at a straw. "Tell me your problem," I said. Out of the corner of my eye I noticed the second ice-cream man laboriously pushing his cart (lid closed, sunshade folded) along the beach towards us. But he was still farther away than our original ice-cream man.

"Its complicated," said Rodney. "If Maxdoch Murwell and I both buy shares, then he'll make $5 million and I'll make $1 million. If we both sell, we'll both break even, no loss, no profit. If I sell and he buys then we both make $4 million. But if I buy and he sells, then he'll make $9 million and I'll *lose* $1 million."

"Magnificent!" I cried. "A classic two-person non–zero-sum game!"

THE PAYOFF MATRIX

In a two-person game, each player has a choice between a finite number of actions or strategies (here, two). The payoff matrix for such a game lists all possible combinations of strategies for the two players, together with their payoffs (how much the players win or lose). The payoff matrix for Maxdoch Murwell and Rodney is:

		Rodney	
		buy	sell
	buy	5,1	4,4
Maxdoch			
	sell	9,−1	0,0

The rows represent Maxdoch Murwell's two strategies, *buy* or *sell*; the columns similarly represent Rodney's strategies. The two numbers in a given row and column show the payoffs, in millions of dollars, to the two players; Maxdoch's payoff is the first entry, Rodney's the second. For example, the lower left-hand entry says that if Maxdoch *sells* and Rodney *buys*, then Maxdoch gains $9 million but Rodney loses $1 million.

"I told you, this is deadly serious —"

"So is game theory," I said. "It was invented by John von Neumann, a top-ranking mathematician. It just sounds frivolous. Let me draw up the payoff matrix" (see the box above entitled "The Payoff Matrix").

"There are various schools of thought about the way game players arrive at a rational set of decisions, which is known as an equilibrium," I said. "One idea is to look for the dominant strategies."

"What are those?" asked Fiona.

"A strategy is just a choice: *buy* or *sell*. As regards being dominant . . ." I wrote in the sand with a stick:

the PAYOFF for a DOMINANT STRATEGY and ANY REPLY
must be STRICTLY GREATER than
the PAYOFF for any OTHER STRATEGY and THE SAME REPLY.

"Strictly greater," I added, "means what it says: greater and not equal. The point is that whatever your opponent does, a dominant strategy gives the greatest payoff."

"Super!"

"If *both* players have a dominant strategy, then their combination is called a dominant strategy equilibrium. The theory is that both players should then play their dominant strategies. Actually that's not always such a good idea, as we'll see, but it often makes sense.

"First, let me see whether *you* have a dominant strategy. It must either be *buy* or *sell*. Suppose you decide to *buy*. Let's consider the outcomes. First, suppose Maxdoch Murwell is going to *buy*:

Rodney *sells*, Maxdoch *buys* payoff to Rodney = $4 million
Rodney *buys*, Maxdoch *buys*: payoff to Rodney = $1 million

and the first choice is better for you. So if you *have* a dominant strategy, it must be to *sell*. Now suppose Maxdoch *sells*:

Rodney *sells*, Maxdoch *sells*: payoff to Rodney = $0 million
Rodney *buys*, Maxdoch *sells*: payoff to Rodney = $−1 million

Again a higher payoff to you if you *sell*. So *sell* is your dominant strategy.

"But does *Maxdoch* also have a dominant strategy? Suppose he decides to *buy*. If you also decide to *buy* then:

Maxdoch *sells*, Rodney *buys*: payoff to Maxdoch = $9 million
Maxdoch *buys*, Rodney *buys*: payoff to Maxdoch = $5 million

and the first choice is better for Maxdoch. But now suppose you *sell*:

Maxdoch *sells*, Rodney *sells*: payoff to Maxdoch = $0 million
Maxdoch *buys*, Rodney *sells*: payoff to Maxdoch = $4 million

and this time the payoff for 'Maxdoch *sells*' is *worse* than for 'Maxdoch *buys*.' So Maxdoch Murwell has *no* dominant strategy.

"And that means there is no dominant strategy equilibrium, so there's no simple rule to make the decision."

"Fat lot of good your game theory is," said Rodney.

DRIVERS' DILEMMA

Horatio Hornblower and Charlotte Clutchgrinder are driving in Paris and both wish to enter an intersection at the same time. Each has two strategies: to beep the horn or remain silent. The payoff, in units of moral supremacy, is 10 units to the one who beeps, but only if the other does not. If both beep, each loses 7 units because the actions become pointless as well as noisy; and if both remain silent the payoff is zero to each.

The payoff matrix is:

		Horatio	
		silent	beep
	silent	0,0	−10,10
Charlotte			
	beep	10,−10	−7,−7

Each player has a unique dominant strategy: *beep*. This leads to a unique dominant strategy equilibrium (*beep, beep*), where Horatio beeps and so does Charlotte. But this dominant strategy equilibrium *loses* them both 7 points! The strategy (*silent, silent*) produces a *better* payoff for both, namely zero. In the jargon, the equilibrium (*silent, silent*) is Pareto efficient.

More traditionally, this game is known as the Prisoner's Dilemma. Here Horatio and Charlotte are two prisoners involved in the same alleged crime who are being interrogated separately. The choices *silent* and *beep* are replaced by *admit nothing* and *confess*; the payoffs represent years in jail (so, for example, −7 is a seven-year jail sentence). The analysis is the same.

"Oh, I haven't finished yet," I said. "I didn't expect there to be a dominant strategy equilibrium, they're not very common. Anyway, the game Drivers' Dilemma (see box) shows that dominant strategy equilibria don't *always* lead to the best outcomes."

As we've just seen, dominant strategy equilibria don't always exist. So we need something different that *does* exist. Another school of thought, developed by John Nash, holds that the correct concept is that of a Nash equilibrium.

"What's that?" asked Rodney.

"Say you choose R and Maxdoch chooses M. It's a choice of strategies for you and Maxdoch such that if Maxdoch sticks to his choice M then YOUR PAYOFF is AT LEAST AS GREAT when you choose R as it is for any other choice, AND if you stick to the choice R than MAXDOCH's PAYOFF is AT LEAST AS GREAT when he chooses M as it is for any other choice. In other words, neither of you has any incentive to change his choice provided the other player doesn't."

"And what's the point of that?"

"A Nash equilibrium," I said, "is a rational strategy for both players and therefore presumably a sensible thing to do."

"Oh."

"That's controversial, of course."

"In economics, *everything* is controversial."

"The nice fact is that Nash equilibria *always* exist. That's known as Nash's Theorem."

"After Nash."

"Quite."

"Like his equilibria."

"Exactly!"

"And his teeth?"

"Shut up, Rodney. Dominant strategy equilibria are always Nash, but Nash equilibria may not be dominant strategy equilibria. It's a more general concept."

"Ah."

"You should bear in mind that there may be several Nash equilibria in a given game, though."

"Oh."

"To work! Let's try $M = buy$, $R = sell$," I said. The payoff to Maxdoch is

$$\text{Maxdoch's PAYOFF } (buy, \, sell) = 4$$

Whereas if he changes his strategy to sell he gets

$$\text{Maxdoch's PAYOFF } (sell, \, sell) = 0$$

which is worse. So he has no incentive to choose differently, provided you don't. On the other hand, for you we have to see what happens if Maxdoch continues to choose *buy* but you change to *buy* also:

$$\text{Rodney's PAYOFF } (buy, \; sell) = 4$$
$$\text{Rodney's PAYOFF } (buy, \; buy) = 1$$

which also means you have no incentive to change either. So (*buy, sell*) is a Nash equilibrium. But not," I added, "A dominant strategy equilibrium, since we've already established that there isn't one."

PROBLEM ❶

Find the Nash equilibrium (or equilibria if several exist) for Drivers' Dilemma.

"Where does all that leave me?" asked Rodney.

"I advise you to sell," I said. "Maxdoch's strategy should then be to buy, and you each make $4 million."

"But what if he sells too?"

"He won't," I said. Now *both* ice-cream men were moving resolutely towards the middle of the beach, farther and farther away from the customers—like us—at the ends. "It wouldn't be rational," I continued. "If he thinks you'll sell, then his best response is to buy—otherwise he gets nothing. I doubt he'd pass up the chance of $4 million profit just to cut your profit to zero!"

"But why *should* he imagine I'm going to sell?"

"Because," I said patiently, "you'd be stupid not to. If you choose *buy*, then your maximum profit is only $1 million. What's worse, Maxdoch could push you a million into the red by choosing *sell*. Would you choose a strategy that gave him that sort of advantage? Of course not! And if we can work that out, so can he, which means he'll know that you'd be an idiot to buy! Then he knows you'll sell—so he has to sell too to maximize his own profit. Simple!"

"Hmmm. I'm not sure Maxdoch Murwell is capable of reasoning that's as convoluted as that."

"No, but I bet his market analyst is."

"True, true . . . and at the worst I still make $1 million." He thought about retrieving his phone, but a hermit crab with rather large claws was living in it, so he walked off to find a public phone. Shortly he returned, looking much happier.

"How about my share?" I asked.

"Tough luck, old boy! You shouldn't have given away your expertise for nothing! Negotiate the price *before* you provide the information!" I saw clearly why Rodney owns a new Lamborghini while I make do with an ancient Lambretta.

Then his face fell. "What are those two idiots up to?"

The two ice-cream men had now settled down side by side in the middle of the beach and were again open for business. Neither looked as if he planned to move again. They now shared the market equally, but the customers had an average walk of a quarter of a kilometer to get ice cream. If the vendors had each set up a quarter of the way along, toward different ends, they'd still have shared the market equally, but the average walk would have been only an eighth of a kilometer. Assuming a uniform distribution of people on the beach, of course.

"That's ridiculous!" said Rodney. "They've come up with the *worst* possible solution!"

"You can't beat market forces, old boy," I said.

PROBLEM ❷

Do you see why the two ice-cream men ended up in the middle of the beach? Can you formulate the problem as a game?

PROBLEM ❸

What is the political analogue of ice-cream men?

PROBLEM ❹ — WORM TURNS

Henry and Anne-Lida Worm are going out for the evening. Independently, they can choose between the tailball match and the opera. They will each arrive separately and there will be no communication between them. Each would prefer to be in the other's company if possible, but Henry prefers the tailball game and Anne-Lida the opera. The payoff matrix is:

		Anne-Lida	
		tailball	opera
	tailball	3,1	−1,−1
Henry			
	opera	−6,−6	1,3

Find all dominant strategy equilibria and Nash equilibria.

PROBLEM ⑤—NUCLEAR DISARMAMENT

A real-world game with a real-world problem! The name is self-explanatory; the payoff matrix is:

		US	
		disarm	arm
	disarm	100,100	−200,200
THEM			
	arm	200,−200	−160,−160

The units are millions of people. Find all dominant strategy equilibria and Nash equilibria.

PROBLEM ⑥—BATTLE OF THE BISMARCK SEA

The scene is the South Pacific in 1943. Admiral Imamura's orders are to transport troops across the Bismarck Sea to New Guinea. Admiral Kenney plans to bomb the troop transports. There are two possible routes, a short northern route and a longer southern one. If Kenney sends his planes the wrong way he can try the other route, but he will lose bombing time. The payoff matrix is:

		Imamura	
		north	south
Kenney	north	2,−2	2,−2
	south	1,−1	3,−3

Find all dominant strategy equilibria and Nash equilibria.

ANSWERS

1. The only Nash equilibrium for Drivers' Dilemma is (*beep, beep*) again. Sometimes rational decisions aren't sensible!

2. This classic example from mathematical economics shows that market forces do not always produce the best result for the customer. Suppose the ice-cream men are Alfredo and Benito, with (say) Alfredo to the left of Benito (Figure 58 *A*). Assume customers go to the nearest ice-cream man. The the dividing-line between their territories is the midpoint M of AB (Figure 58 *B*). If Alfredo is not exactly at the center of the beach, then Benito can improve his market share by moving *between* Alfredo and the center (Figure 58 *C*). Similarly, if Benito is not at the exact center, then Alfredo can do the same. Therefore both ice-cream men will move to the center. Each gets half the market (Figure 58 *D*).

However, they have now managed to *maximize* the average distance that customers have to go to reach them. If instead they agree to move to 1/4 and 3/4 of the way along the beach, respectively (Figure 58 *E*), then each still retains half the market, but the distance customers have to walk is minimized. But there is no incentive for the vendors to choose (1/4, 3/4).

This is rather like Drivers' Dilemma, except that each player has infinitely many possible strategies (distances along the coast). To simplify the representation as a game, suppose that Alfredo can choose either to move to the quarter mark or the middle, and Benito to the three-quarter mark or the middle. The payoff matrix *in terms of the vendors' market share* is then:

Benito
3/4   1/2
1/4   1/2, 1/2   3/8, 5/8
Alfredo
1/2   5/8, 3/8   1/2, 1/2

The only Nash equilibrium is (1/2,1/2), and this is also a dominant strategy equilibrium.

But now consider the game from the point of view of the customers. Now the payoff is the average distance to the nearest vendor, to which we assign negative values because smaller distances are better. The payoff matrix becomes:

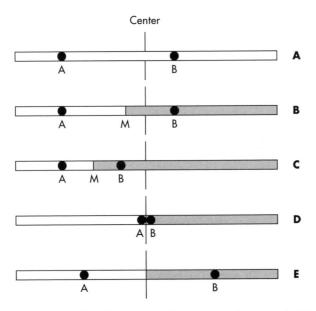

FIGURE 58   Two ice-cream salesmen on a beach and their market shares.
A. Arbitrary starting position. *B.* Corresponding market shares. *C.* Gaining market share by relocating premises. *D.* The end result of market forces. *E.* A solution not produced by market forces but better for the customers and just as good for the vendors.

	Benito's customers	
	3/4	1/2
1/4	−1/8,−1/8	−5/48,−17/80
Alfredo's customers		
1/2	−17/80,−5/48	−1/4,−1/4

Now the Nash equilibrium (and dominant strategy equilibrium) is (1/4,3/4). The customer is always right!

3. Replace the beach by a linear scale of political ideology, from left to right. The game then models a two-party system in which, in order to woo the maximum number of voters, both parties try to "borrow each others' clothes" and end up at the political center. This has been happening in the United Kingdom over the past decade. It seems to have happened long ago in the United States.

4. There is no dominant strategy equilibrium for Worm Turns. Both (*tailball, tailball*) and (*opera, opera*) are Nash equilibria.

5. Nuclear Disarmament, of course, is Drivers' Dilemma in disguise. The dominant strategy equilibrium is (*arm, arm*), and this is also the unique Nash equilibrium.

6. In Battle of the Bismarck Sea neither player has a dominant strategy, so there is not dominant strategy equilibrium. Kenney will choose *north* if he thinks Imamura will choose *north*, but *south* if he thinks Imamura will choose *south*. Similarly, Imamura would always prefer to choose the opposite of what Kenney chooses.

   Imamura does have a *weakly* dominant strategy (remove the adjective "strictly" from "strictly greater" in the definition). Namely, Imamura never does any worse by choosing *north*; and if Kenney happens to choose *south* then Imamura does better.

   If Kenney realizes this, he can assume Imamura will go north and cross off the second column to create a new game. Now Kenney has a dominant strategy, namely *north*. The combination (*north, north*) is called an iterated dominant strategy equilibrium. It's also the unique Nash equilibrium. It's also what actually happened.

FURTHER READING

Nash, John. Equilibrium points in *n*-person games. *Proceedings of the National Academy of Sciences* 36 (1950): 48−49.

————. Non co-operative games. *Annals of Mathematics* 54 (1951): 286–95.

Rasmusen, Eric. *Games and Information.* Oxford and New York: Basil Blackwell, 1989.

Thomas, L. C. *Games, Theory and Applications.* Chichester, England: Ellis Horwood, 1984.

Von Neumann, John, and Oskar Morgenstern. *Theory of Games and Economic Behavior.* Princeton: Princeton University Press, 1947.

10

Curie's

Mistake

A few summers back I was driving along a freeway in upstate New York. Ahead was a large truck, with two mud-flaps at the rear. They were flapping, as all good mud-flaps should. But they weren't flapping in unison. When the left-hand flap was moving forward, the right-hand one was moving backward, and vice versa. An engineer would have noted that the oscillations were 180° out of phase. A physicist would have observed that the oscillations were caused by vortex-shedding: the truck was leaving a train of tiny tornadoes in its wake, peeling off in turn to the left and the right, and wiggling the flaps as they passed. But what I saw was an example of symmetry-breaking. The arrangement of flaps on the truck was, near enough, left-right symmetric, but the motion was asymmetric: the left-hand flap and the right-hand flap were not performing identical motions. In fact, the pattern of vortices (Figure 59A) has its own symmetry, but of a different kind from that of the truck. The truck is symmetric under a

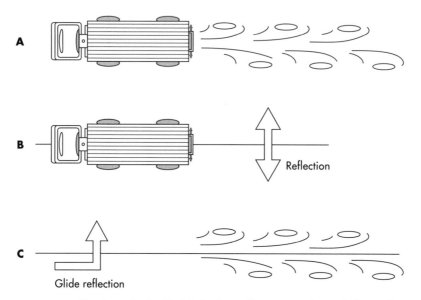

FIGURE 59   Vortices *A.* shed behind a laterally symmetric truck do not have the truck's bilateral symmetry *B.* but instead are symmetric under a glide reflection *C.* As a result the mud-flaps waggle alternately rather than in unison.

reflection that interchanges left and right (Figure 59*B*); the vortex train that it sheds is symmetric under a glide reflection (Figure 59*C*).

On an earlier occasion I was in northern California, where the huge redwoods and sequoias grow. The trunk of a tree is approximately cylindrical, and it thus has a good approximation to cylindrical symmetry. The symmetries of a cylinder are of three kinds: rotations, translations, and reflections. If you rotate a cylinder about its axis (Figure 60*A*) it looks exactly the same as before; and the same is true if you translate it in the direction of its axis (Figure 60*B*). To be precise, translational symmetry holds only for an infinitely long cylinder, but it is valid to a good approximation for a sufficiently long one. There are also two distinct types of reflectional symmetry (Figures 60*C* and *D*), vertical and horizontal.

It seems plausible that the pattern of bark on the tree should have symmetry similar to that of the tree itself. Now a pattern of bark with a good approximation to full cylindrical symmetry will have to look pretty much the same after any of those rotations, translations, and reflections.

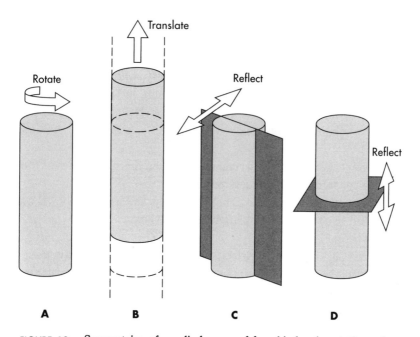

FIGURE **60**   Symmetries of a cylinder are of four kinds: *A*. rotations about the axis; *B*. translations (assuming an infinite cylinder); *C*. reflection in a plane through the axis; *D*. reflection in a plane perpendicular to the axis.

That means that the grooves in the bark should run roughly vertically, as they do on most trees. But what I actually saw, on *some* of the Californian trees, was a *spiral* pattern on the bark, winding round like stripes on a barber's pole or on sugar candy. The spiral still has some symmetry, but of a different kind. If you rotate a helical spiral, *and* translate it along its axis, it looks the same. So the symmetry of a spiral is a mixture of rotation and translation, known as a *screw*. Indeed, this is why a carpenter's screw works: as it rotates *and* goes deeper into the wood — that is, as it translates — it fits into its own hole. Real woodscrews are tapered so they enlarge the hole slightly as they go, for a tight fit; but a bolt with a helical thread has exact symmetry of this kind.

Did something strange happen to those trees that developed spiral bark? Pesticides, a bad winter, drought? Or should we *expect* spiral patterns as well as perfectly symmetric ones? To approach such problems we must answer a fundamental general question: how does the symmetry of a system affect its behavior?

A famous answer was given by the great physicist Pierre Curie, who is best remembered for his work on radioactivity, with his wife, Marie, which led to the discovery of radium and polonium. In 1894 Pierre Curie gave two logically equivalent statements of a general principle from the folklore of mathematical physics:

1. If certain causes produce certain effects, then the symmetries of the causes reappear in the effects produced.
2. If certain effects reveal a certain asymmetry, this asymmetry will be reflected in the causes which give rise to them.

But was Curie correct? Curie's answer is often assumed by scientists, and often unconsciously, in the course of their work. For example, in the Kensington Science Museum in London there is an engineering model of a passenger jet, used in a wind tunnel to study the flow of air around the aircraft. Since the aircraft is bilaterally symmetric, the engineers built only half of the model — tacitly assuming that the air flow had to be bilaterally symmetric as well. Are such assumptions justified?

At first sight, Curie's principles are "obviously" true. If a perfectly spherical planet acquires an ocean, we expect the ocean to have the same depth everywhere. If the planet rotates, destroying its spherical symmetry but retaining rotational symmetry about an axis, then we expect the ocean to bulge at the equator but retain circular symmetry. What else could the airflow be about a bilaterally symmetric aircraft *except* bilaterally symmetric?

That wasn't a rhetorical question. It has an answer, and a surprising one: the flow could be *asymmetric*. Indeed, the mud-flaps of the truck and the bark of the tree described above are examples of symmetric systems whose behavior is *less* symmetric. There are many others. If a perfect circular cylinder, say a tubular metal strut, is compressed by a sufficiently large force, it will buckle. The buckling is *not* a consequence of lack of symmetry caused by the force: even if the force is directed along the axis of the tube, preserving the rotational symmetry about that axis, the tube will still buckle. Buckled cylinders cease to be cylindrical — that's what "buckle" means. Buckled spheres cease to be spherical. A computer picture of a spherical shell buckled by a spherically symmetric compressive force is shown in Figure 61: the symmetry breaks from spherical to circular.

This phenomenon, whereby symmetric systems produce behavior with less symmetry (or none) is called symmetry-breaking. It seems to be responsible for many types of pattern-formation in nature. It also has a very well defined mathematical structure that can be used to understand the processes that give rise to patterns.

What causes the symmetry to break? The answer is that natural systems must be stable; that is, they should retain their form even if they are disturbed. A pin lying on its side is stable and can occur in the real world. A pin balanced on end is *theoretically* possible — it is a valid solution of the mathematical equations that describe how a pin should behave — but it does not occur in the real world, because it is unstable: it will topple at the slightest breath of wind. Mathematically speaking, a long thread or chain can also be balanced on end; but that doesn't explain the Indian rope trick, again because the rope is unstable. Curie is right in asserting that symmetric systems should possess symmetric states; but he failed to address their stability. If a symmetric state becomes unstable, then the system will do something else — which need not be symmetric.

In particular, there's nothing surprising about trees with spiral patterns to their bark. If the perfectly symmetric pattern represents an

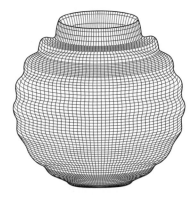

FIGURE 61   Computer picture of a spherical shell buckled by a uniform spherically symmetric force. The buckled state has circular, but not spherical, symmetry.

unstable development, then tiny disturbances will cause the symmetry to break. Spirals are one of the common ways to break cylindrical symmetry, so a spiral pattern might develop instead.

PROBLEM ❶

Can you think of five naturally occurring examples of symmetry-breaking?

You can experiment on symmetry-breaking in your own home. Take a hose of circular cross-section and hang it vertically, nozzle downward, with water flowing steadily through it. You could try it with flexible rubber tubing of the kind found in chemistry laboratories, about 5 mm in diameter. This system is circularly symmetric about an axis running vertically along the center of the hose. And indeed, if the speed of the water is slow enough, the hose just remains in this vertical position, retaining its circular symmetry.

However, if the tap is turned on further, the hose can begin to wobble. In fact, there are two distinct kinds of wobble; which one occurs in your experiment will depend on the length and flexibility of the tubing. In one, the hose swings from side to side like a pendulum (Figure 62A). In the other, it goes round and round, spraying water in a spiral (Figure 62B). Similar effects are often observed when children wash the family car. These wobbles do not possess circular symmetry about a vertical axis: indeed, they break it in two distinct ways.

They also break a less obvious but very important symmetry: symmetry in time. The original steady flow looks exactly the same at all instants of time; the oscillating flows do not. The time symmetry is not totally lost, however: both wobbles are periodic, hence look exactly the same when viewed at times that are whole-number multiples of the period. The continuous temporal symmetry of a steady state breaks to give the discrete symmetry of a periodic one (Figure 63).

Symmetry-breaking leads to the formation of patterns. In some parts of the world there are curious flat mounds of stone, arranged in a roughly hexagonal pattern like a honeycomb. Why? Originally the stones were on the bed of a large, shallow lake. The sun's rays heated the lake, giving rise to currents in the water. Now a large lake is approximately symmetric under translations in any direction, as well as rotations. If no

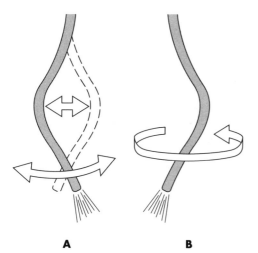

A   **B**

FIGURE **62**   Two ways for a circular hose to wobble: *A.* a pendulum-like standing wave; *B.* a rotating rave. Neither wobble has the full cylindrical symmetry of the hose.

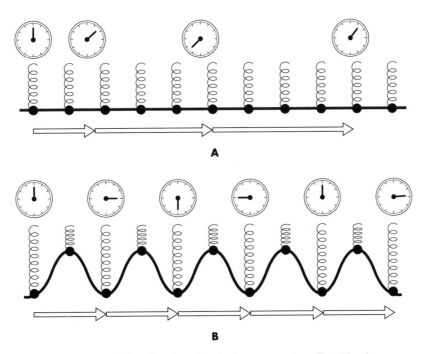

A

B

FIGURE **63**   All periodic vibrations break time symmetry. The idea is illustrated here for a weight hanging on a spring. *A.* The system hangs in equilibrium, and is symmetric under time-translation: at any subsequent time it looks exactly the same. *B.* When the weight oscillates periodically, the system only looks the same after times that are whole-number multiples of the period.

symmetry were broken, the flow of water would also be symmetric under all translations and rotations—which means no flow at all! All that would happen is that the heat would be *conducted* through a stationary lake.

To find out what really happens, take a frying pan filled with a shallow layer of water (the lake) and sit it on the stove (the sun). Turn on the stove and heat the pan. (Be careful, and if you are under the age of twenty-one or married to a cook ask permission first.) What happens?

You get strange cellular patterns.

The physical reason is that hot fluid tends to rise. As the temperature increases, the hot layer on the bottom is trapped beneath a layer of colder, denser fluid. This situation is unstable and is destroyed by the onset of convection. Hot fluid rises in some regions, cold fluid descends in others. A cellular pattern of moving fluid, known as Bénard cells, appears. Sometimes the pattern consists of parallel rolls, sometimes it is a honeycomb array of hexagons. In a real pan the symmetry is approximate and the pattern is rather irregular, but in an idealized infinite pan you get perfect honeycombs. These patterns still have a great deal of symmetry, but less than that of the pan.

The truth is thus more complex than Curie's principles suggest. (To be fair, Curie knew this, as he elsewhere made clear.) A symmetric system will take up an equally symmetric state, *except* when it doesn't! It might seem that this just says that Curie was right except when he was wrong, but by analyzing the conditions under which symmetry-breaking occurs we can give the idea some genuine content. To do that, we need to make the idea of symmetry precise, and that requires the mathematical concept of a group.

Symmetry is basic to our scientific understanding of the universe. Conservation principles, such as those for energy or momentum, express the symmetry of space–time: the laws of physics should be the same everywhere. The quantum mechanics of fundamental particles, a crazy world in which a proton can be "rotated" into a neutron and whose laws must reflect this possibility, is couched in the mathematical language of symmetries. The symmetries of crystals not only classify their shapes but also determine many of their properties. Many natural forms, from starfish to raindrops, from viruses to galaxies, have striking symmetries. Man-made articles tend to be symmetric: cylindrical pipes, circular plates, square boxes, spherical bowls, hexagonal steel bars.

"What immortal hand or eye/could frame thy fearful symmetry," asked
William Blake, referring to the "tyger."

But what, exactly, *is* symmetry?

In everyday language the term "symmetry" is used in two distinct
ways. The first is rather vague, something along the lines of the elegant
proportions of Blake's tiger. The second is more specific, referring to a
repetitive feature of a shape. It is this second meaning that interests
mathematicians.

The human form is (approximately) bilaterally symmetric: people
seen in a mirror look much the same as they do in the flesh. That is, the
left-hand side and the right-hand side agree in their general outlines
(Figure 64*A*). A starfish (Figure 64*B*) has fivefold symmetry: each of its

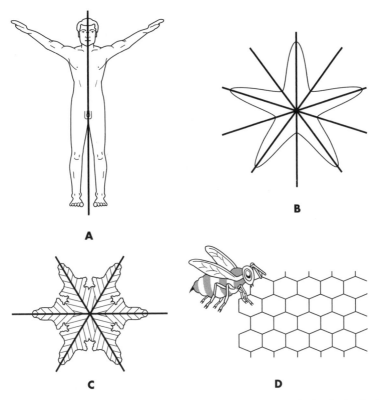

FIGURE **64**   Different kinds of symmetry: *A*. bilateral; *B*. fivefold; *C*. sixfold;
D. sixfold, with extra translations.

five arms is the same shape as the others. A snowflake (Figure 64*C*) has sixfold symmetry, and an infinite honeycomb (Figure 64*D*) has a spatially extended repetitive structure in addition to the sixfold symmetry of each cell.

In order to capture the essence of symmetry in this second sense, mathematicians focus not so much on the shape of the object, but on the transformations that may be applied to it. Suppose that someone is shown a perfectly symmetric starfish placed on a table-top, and then, while that person looks the other way, the starfish is rotated by one fifth of a turn. On taking another look at the starfish, it will be impossible to decide whether or not it has been moved. The same is true if it is rotated through two fifths of a turn, or three fifths, or four fifths; or indeed if it is simply left undisturbed. There are thus five distinct transformations that may be applied to the starfish which leave its apparent form and position unchanged. These transformations between them determine the symmetry group of the starfish.

The word "group" here does not just mean that there are several transformations. Suppose that two successive symmetry transformations are applied. Each leaves the starfish apparently unchanged, so the final result also leaves it apparently unchanged. That is, the result of performing two successive symmetry transformations must be yet another symmetry transformation. The word "group" expresses this fact, that any two symmetry transformations, when combined together, yield another one.

For example, rotating the starfish one fifth of a turn and then two fifths of a turn has the same effect as rotating it three fifths of a turn. Symbolically this can be written (in units of "one turn") as $\frac{1}{5} + \frac{2}{5} = \frac{3}{5}$, a natural enough equation. The mathematics of symmetry groups is not quite so simple as this, however. Imagine rotating three fifths of a turn and then two fifths. The result is a full turn; but that really does leave every point of the starfish in exactly the same place that it started. If we concentrate only on where points end up, and not on how they got there, this is the same as "no rotation." In other words, $\frac{3}{5} + \frac{2}{5} = 0$ in the world of starfish symmetries!

The most important types of symmetry transformation are rotations, which leave some point — the center of rotation — fixed; reflections, which effectively view the shape under discussion in an imaginary mirror; and translations, which move the shape bodily in some direction without rotating or reflecting it. The word "transformation" is normally suppressed: one refers just to the "symmetries" of the object. Thus in group-theoretic parlance a square has eight symmetries: three rotations (through a quarter, a half, or three quarters of a turn), four reflections (in the two diagonals and the two lines dividing opposite sides in half), and the trivial symmetry "do nothing." A single square has no translational symmetries. On the other hand an infinite plane covered in square tiles also has translational symmetries: if the entire pattern is moved sideways a whole number of tiles, it looks exactly the same. This may seem a fanciful example, but it is essentially a two-dimensional version of the way physicists capture the symmetry of crystals.

Where does symmetry go when it breaks? A good question! The catastrophe machine (Figure 65), invented for rather different reasons

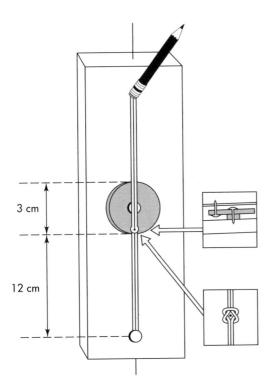

FIGURE 65   The Zeeman catastrophe machine and its axis of symmetry. See text for instructions.

by Christopher Zeeman of Warwick University in 1969, shows that symmetry is not so much *broken* as *spread around*. You can make one and experiment with it. By following the instructions given below you can make a workable model; if you want something more permanent you have to do a more thorough engineering job.

Attach a circular disk of thick card, of radius 3 cm, to a board using a drawing pin and a paper washer. Fix another drawing pin near the rim of the disk with its point upward. To this pin attach two elastic bands, of about 6 cm unstretched length. Fix one to a point 12 cm from the center of the disk, and leave the end of the other free to move along the center line as shown, for example by taping it to a pencil which you can move by hand.

The entire system has reflectional symmetry about the center-line. And if you begin to stretch the free elastic, then you'll find that the system obeys Curie's principles and stays symmetric — the disk does not rotate (Figure 66A). But as you stretch the elastic further, the disk suddenly begins to turn — maybe clockwise, maybe counterclockwise (Figure 66B). Now the state of the system fails to have reflectional symmetry. The symmetry has broken, and Curie's principles have failed.

What has happened to the missing symmetry?

Hold the elastic steady and rotate the disk to the symmetrically placed position on the other side (Figure 66C). You'll find that it will remain there. Instead of a single symmetric state we have two symmetrically related states. In general, a system with a given symmetry group breaks symmetry to a smaller group, or subgroup. Moreover, the system can then exist in several states, each obtainable from the others by one of the symmetries of the full system.

For example, the buckled spherical shell (Figure 61) breaks symmetry from spherical to circular, from the group of all rotations in three-dimensional space to its subgroup of rotations with a given axis, here visible as the axis of symmetry of the buckled sphere. But the symmetry is not lost completely: if you rotate the buckled sphere in any way whatsoever you get another possible way for the sphere to buckle. In principle, infinitely many distinct axes can occur; in practice, imperfections in the shell select one of them.

Back to Blake's tiger. Let's apply general principles of symmetry-breaking to the patterns on the tiger's skin. To a first approximation, tigers are cylindrical. We've seen one possible way for cylindrical symmetry to break: to spirals. Another type of symmetry-breaking occurs when the translational symmetry is broken. Then a uniform pattern is replaced by periodic stripes (Figure 67). Did Blake stumble on a deeper truth than he imagined?

How might this come about in nature? Suppose that the patterns on a mature tiger's skin are controlled by chemicals which diffuse over the surface of an embryonic tiger as it grows. If we model the tiger by a perfect cylinder then a fully symmetrical pattern of pigmentation would

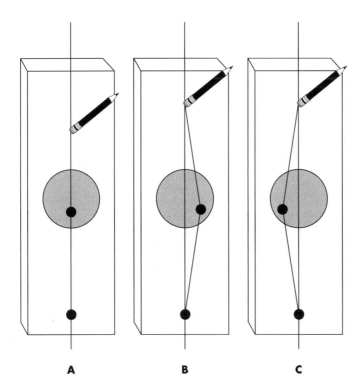

A   B   C

FIGURE 66   *A.* If the free elastic is stretched slightly then the catastrophe machine takes up a symmetric position, on the axis. *B.* If the elastic is stretched further, then despite the bilateral symmetry, the disc rotates off the axis. However, there is a second, symmetrically related state (*C*). The individual solutions (*B, C*) break the symmetry, but together they form a symmetric pair. Thus the symmetry is spread over several solutions rather than concentrated on one.

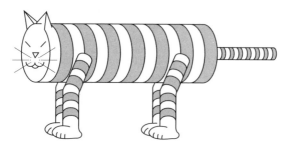

FIGURE **67**   Cylindrical approximation to a tiger with symmetry-breaking stripes.

give a uniformly orange tiger—in other words, a lion. But a uniform distribution of chemicals can be unstable. Then the symmetry breaks—and one possibility is stripes! Is this the main difference between lions and tigers?

This idea goes back to Alan Turing, one of the fathers of the computer, who also took an interest in biomathematics. Jim Murray of Oxford University has used a computer to model the same phenomenon on a more realistic approximation to a whole animal, with similar results. Indeed, the idea can be pushed further. There is a second instability, whereby the stripes themselves break up into spots arranged in a hexagonal pattern. Now the tiger becomes a leopard. Computer calculations of the patterns on big cats' tails are shown in Figure 68. Long thin stripes turn out to be less stable than short fat ones; and an animal's tail is thinner than its body, so stripes on the tail are shorter than those on the body. They thus break up less easily. So a spotted animal can have a striped tail but a striped animal cannot have a spotted tail.

There are still many unsolved problems. For example, what prevents the occurrence of spigers—tigers with spiral stripes? Or squeopards—leopards whose spots are arranged in squares rather than hexagons? But the idea of symmetry-breaking illuminates a remarkable variety of scientific disciplines, from astronomy to zoology.

We've seen the zoology. For astronomy, we need look no further than a remarkable recent discovery known as Godfrey's kinky current. The planet Saturn has rotational symmetry about its axis, and the wind-patterns on its surface create colored circular bands parallel to the equator. (Saturn is striped, too: it's a planetary "sphiger.") But in 1988

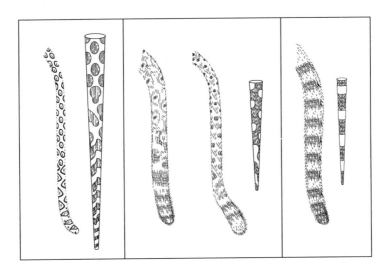

FIGURE **68**   Computer model and reality for the tails of the leopard (*left*), jaguar or cheetah (*center*), and genet (*right*). [James Murray, How the leopard gets its spots, *Scientific American* 258 (March 1988): 62–69.]

D. A. Godfrey, analyzing *Voyager* pictures by computer, discovered that around the north pole of Saturn there is a slowly rotating *hexagon*! The circle's symmetry under rotations through arbitrary angles has been reduced by rotations through multiples of 60°: circular symmetry has broken to hexagonal symmetry!

Not everything in the world is symmetric, and not every symmetric system can profitably be analyzed in terms of symmetry-breaking. But a surprising variety can. Unlike most things in the world, symmetries are most effective when they are broken.

ANSWERS

1. There are literally thousands of examples of symmetry-breaking in nature. Here are just a few:

   Spiral galaxies. A disc of matter breaks its circular symmetry to form equally spaced spirals.

Drops of dew on a spider's web. A uniform distribution of water along the web, like a thin coat, has translational symmetry but is unstable. It breaks up into periodically spaced droplets.

The jet stream in the earth's upper atmosphere. The earth has circular symmetry, but the jet stream develops large, roughly equally spaced waves.

The vibration of a clarinet reed. The clarinetist blows steadily and uniformly, but the reed vibrates to and fro.

A circular drop of water in a very hot pan will vibrate in polygonal patterns, breaking circular symmetry, as steam builds up beneath it.

A spherical balloon, blown up too far, distorts into a non-spherical shape. Or bursts!

The universe. Spherical at the time of the big bang, but now filled with an endless variety of forms; perhaps still spherical, but certainly no longer spherically *symmetric*.

FURTHER READING

Alperin, Jonathan. Groups and symmetry. In *Mathematics Today*, ed. Lynn Arthur Steen. New York: Springer-Verlag, 1978.

Gierasch, Peter. Hexagonal polar current on Saturn. *Nature* 337 (26 January 1989): 309.

Golubitsky, Martin, and Michael J. Field. *Symmetry in Chaos.* Forthcoming.

Golubitsky, Martin, and Ian Stewart. *Fearful Symmetry—Is God a Geometer?* Oxford and New York: Basil Blackwell, forthcoming.

Golubitsky, M., Ian Stewart, and David G. Schaeffer. *Singularities and Groups in Bifurcation Theory.* Vol. 2. New York: Springer-Verlag, 1988.

Krantz, William, Kevin Gleason, and Nelson Caine. Les sols polygonaux. *Pour la Science* 136 (February 1989): 18–24.

Murray, James. How the leopard gets its spots. *Scientific American* 258 (March 1988): 62–69.

Weyl, Hermann. *Symmetry.* Princeton: Princeton University Press, 1952.

A Dicey
Business

———

"**T**wenty-six," said Bumps the goose-girl, giving the die a quarter turn so that face 5 was on the top.

Grimes the shepherd-boy grimaced. "You rotten pig! I *need* that 5 to win!"

"That's why I put it on top, so you can't use it," said Bumps. "First person over 31 loses. You can only play 1, 3, 4, or 6 with a quarter turn. Playing 6 makes you lose straight away. If you play 1 the total goes to 27, but then I play 4 and win. If you play 3 the total goes to 29 and I play 2 and win. If you play 4 the total goes to 30 and I play 1 and win. So I win!"

I suppose you want to know the rules. The first player places an ordinary cubical die, with faces marked 1–6, on the table. The score is given by its top face. Players alternately roll the die through a quarter turn to bring a new, adjacent face to the top, adding the value of the top face to the total. The first to make the total exceed 31 (or some other agreed figure) *loses*. See Figure 69 for a sample game.

Grimes sighed and rolled on to his back, chewing at a straw. "You've beaten me again," he said sadly.

"I told you," said Bumps. "I'm a perfect logician. I knew I'd win from the opening move—you didn't stand a chance. Do you want another game?"

"No. What's logic got to do with winning games? I usually beat you at tennis."

"That's a physical game, and you're stronger than I am. But this is a game of pure intellect, so I win hands down, because I know the perfect strategy."

"Strategy . . . people use that word a lot. But I've never understood quite what it means. I suppose it's how to play the game well."

"In this sort of game it's how to *win*, no matter what your opponent does," said Bumps. "Some games have a winning strategy for the first player, some for the second. Some games, like tennis perhaps, don't have perfect strategies at all."

"How do you tell which is which?"

"Logic. Which also lets you find a strategy if there *is* one."

"I don't follow you."

Bumps stretched her legs out in the long grass and smoothed her skirts to cover up her ankles again. "Let me put it this way. Take any game that satisfies four simple rules:

1. There is only a finite number of possible positions.
2. Any play of the game ends after finitely many moves.

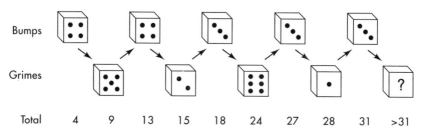

FIGURE **69** Sample game: Grimes loses, as usual.

3. The game always ends in a win for one player (no draws).

4. The possible moves from a given position are the same for both players.

PROBLEM ❶

A. *Convince yourself that the game Bumps and Grimes play satisfies all four rules. "Positions" in that game consist of two numbers: the face of the die that is showing and the running total.*

B. *Which rules does chess satisfy?*

"Now," said Bumps, "any such game either has a perfect winning strategy for the first player, or it has one for the second player. And I'll tell you *how* to win any game satisfying my rules. The way to win such a game is always to start from a winning position. If you have to start from a losing position, then you lose. By which I mean that the *other* player has a winning strategy. Of course, he may be too stupid to play it, but we perfect logicians—"

"Yes, but—I mean, how can you tell—"

"A winning position is one in which you can choose *some* move that leads to a losing position (for your opponent). A losing position is one for which *every* possible move leads to a winning position (for your opponent). Oh, yes, one final point: the end positions have to be defined either as winning positions or losing positions as part of the rules of the game. For instance, in our game, any position where the running total is *exactly* 31 is defined as a winning position."

"That's silly," said Grimes. "First you define 'winning position' in terms of 'losing position' and then you define 'losing position' in terms of 'winning position.' Logic? Rubbish! You're arguing in circles!"

"No I'm not," said Bumps, "though I admit it looks as if I am. Actually, I'm using recursion. If you take any position and follow through the logical chain, you'll find that eventually you get to an end position, which we know is either a win or a loss, and then you can unravel the chain of logic to find out what the original position is. In practice, since that ends up by making you work backward, you may as well start out by working backward in the first place."

"Eh?"

"Let me give you an example."

From the pocket of her smock she pulled out a bar of chocolate (Figure 70). "The moldy square of chocolate has been doctored to make it taste — and look — revolting. The rules of the game are very simple. Players take turns breaking off pieces of chocolate by making a single straight cut, along one of the grid lines, right across the bar. They eat the piece that they break off. Whoever eats the moldy piece loses. A position is specified by the sizes of the sides of the bar. This one starts out in position 6×10. You can go first. What move do you make?"

"Um. That moldy bit looks pretty dreadful . . ."

"As I said, work backward. If you're left with just the moldy piece on its own, that's a losing position. So any position leading to that, in one move, is necessarily a winning position. And those positions are the ones that leave a strip of chocolate, of size 1×2 or 1×3 or 1×4 or in general $1 \times n$ for $n > 1$. See that? For instance, if you've got a piece that's size 1×5 then you break off everything *except* the moldy bit [Figure 71A]. Moreover, *only* the positions $1 \times n$ can lead to 1×1 in a single move."

"What about 2×1, 3×1, and so on?"

"They're effectively the same as 1×2 and 1×3 — I'm not distinguishing between the two possible orientations of a given shape."

"Oh. So I — wait a bit, I'm looking at 6×10. That's not $1 \times$ anything. Does that mean I lose? Have you beaten me again, even before I make a move?"

"Maybe, but we don't know that yet. We haven't worked backward far enough. Let's take the next step. Any position that *always* leads to a winning position must be a losing position. Can you think of any position that always leads to something of the shape $1 \times n$, $n > 1$?"

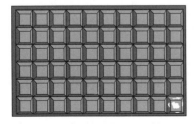

FIGURE **70** Don't eat the moldy piece of chocolate!

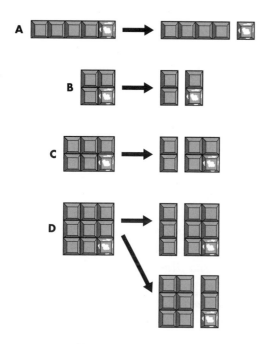

FIGURE 71   Reconstructing the strategy by working backwards.

Grimes thought hard. "No."

"2 × 2."

"Why?"

"The only possible move is to break it in half [Figure 71*B*], leaving 1 × 2."

"Oh, right!"

"I can't see any other losing moves at this stage, though. So now we take another look for winning positions. The only new ones will be those that lead to 2 × 2 in a single move. What are they?" "Um . . . 2 × 3, because I can break off a 2 × 1 piece [Figure 71*C*]. And 2 × 4, 2 × 5, . . ."

"Exactly! In general, 2 × *n* where *n* > 2. Now what?"

"We look for a new losing position, always leading to positions we already know are winning . . . I see it! 3 × 3!" (Figure 71*D*).

"Exactly. And then we find new winning positions that lead to that, namely 3 × 4, 3 × 5, . . . and in general 3 × *n* where *n* > 3."

"I'm starting to see a pattern."

"So am I."

PROBLEM ❷

What is the pattern, and can Grimes win the 6 × 10 game?

PROBLEM ❸

In Divide and Conquer, players start with two boxes, each containing a non-zero number of counters. Taking turns, they throw away the contents of one box and divide the contents of the other between the two boxes, each with a non-zero number. The game ends when a player cannot make a move that obeys these rules, and this player then loses. Which positions are winning positions and which are losing positions?

"The die-turning game can be tackled in the same way," said Bumps. "There's nothing special about the total 31, either, so I'll show you how to find a winning strategy for any selected total.

"As you've seen, the position in the game is most obviously given by two numbers, the face that is uppermost and the current total. But I'm going to change those to simplify the working.

"First, instead of the current total I'm going to use the *difference* between the current total and the agreed limit. That is, the number of points that *remain* before a player exceeds the limit. It's easier to work backward if we use that.

"Second, observe that the available moves when face 1 is uppermost, namely 2, 3, 4, 5, are the same as when face 6 is uppermost. The moves from *opposite* faces form a band between them. So a position in which face 1 is on top is effectively the same as one for which face 6 is on top. The same goes for faces 2 and 5, or 3 and 4. So there are really only *three* face positions: 1/6, 2/5, or 3/4."

"Hang on. If I move to face 6 the total changes by 6, but if I move to face 1 it only changes by 1."

"Ah, but those are the *moves*, not the *positions*. I agree that we have to distinguish between opposite faces for the moves."

"Oh."

"Working backward to find the winning and losing positions is complicated, and it's easy to make mistakes. So I'm going to make myself a small analog computer." She extracted a pair of scissors and cut out three cards, which looked like Figure 72A–C. Then she drew three columns on a piece of paper, as in Figure 72D.

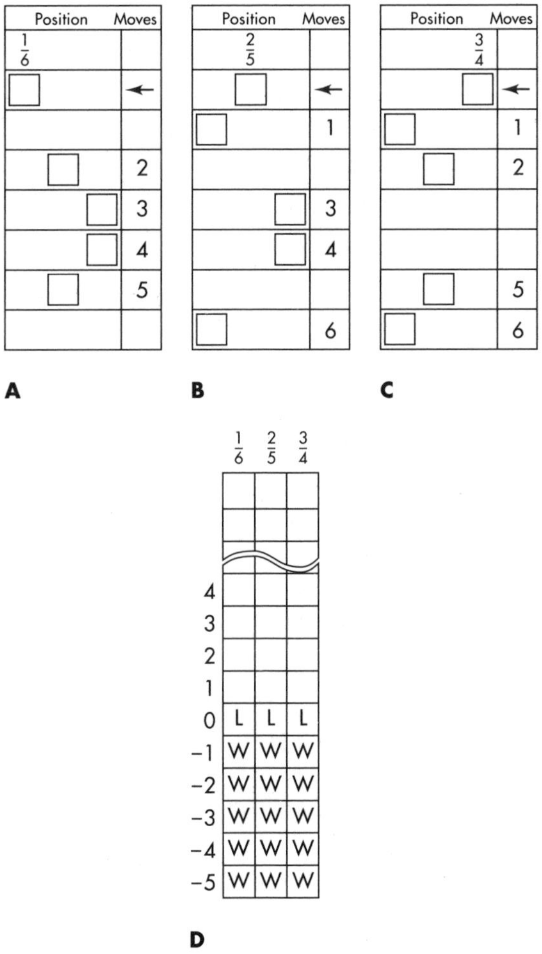

A

B

C

D

FIGURE 72   *A–C.* Three strategy cards: cut windows where the small squares are. *D.* The initial grid.

"The grid represents the possible positions in the game. The columns show the top face of the die; the rows show the remaining total. I've filled in a row of Ls to show the losing positions when the total left is zero."

"What about all those Ws?"

"Those are a convention that makes it easier to get the calculation started. Let's agree that if you face a negative total you must have won, because the previous player must already have lost. So negative totals should be thought of as winning positions."

"Well . . ."

"Watch how it goes and you'll soon see that it makes sense," said Bumps firmly. "I haven't explained the cards yet. There's one for each possible top face 1/6, 2/5, or 3/4. Take the card for 1/6. The hole immediately below the 1/6 is there so that you can write on the grid. The other four holes show how the four possible moves 2, 3, 4, or 5 change the total that remains. Look, if you play 2 then the total reduces by 2, so the corresponding hole is 2 lines farther down, and so on."

"I think I see."

"I can use the cards to work backward. To find out the winning moves in a given row I place the three cards in turn on the grid, so that the writing holes, indicated by arrows, line up with that row. If *no* Ls appear in any of the other holes, that means every move I can make leads to a winning position. That's the definition of a *losing* position, so I write an L in the arrowed hole. On the other hand, if Ls *do* appear, then I have a winning position, so I write in the corresponding moves instead. Look, I'll show you how to get rows 1 and 2 [Figure 73*A* and *B*].

"By repeating this you can work steadily upward to higher and higher totals [Figure 74]. Now you've found the perfect strategy! For a given remaining total and top face you look in the corresponding row and column. If the entry is L, you've lost against a perfect player — who will use the same grid to make sure she wins! If the entry isn't an L, then you can make any of the moves listed in that cell, safe in the knowledge that they're all winning moves.

"There's one final complication which you must bear in mind. The *first* move is arbitrary. Starting with a total of 31, for example, I can create the following positions" — Bumps quickly wrote out a table — "You *could* use the grid to work out which are wins or losses:

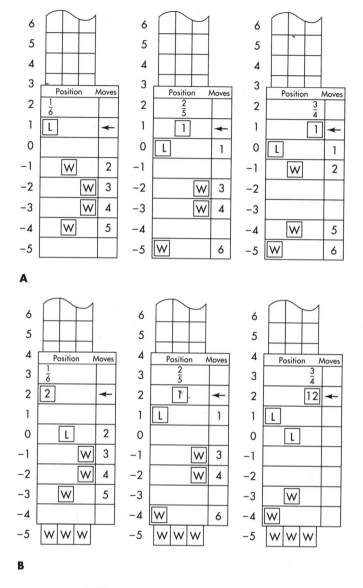

A

B

FIGURE **73** *A.* How to find out which strategy to use in row 1. Place the card for position 1/6 so that the arrow is in row 1 and look at the windows below. Every window shows a W, so the position is a losing one: write L in the window opposite the arrow. Repeat with card 2/5. This time there is an L in the window opposite move 1, so 1 is a winning move: write 1 in the window opposite the arrow. Finally repeat with card 3/4: again, 1 is a winning move. *B.* Now move on to the second row, and so on, working backward up the grid as far as you wish.

Top Face

	$\frac{1}{6}$	$\frac{2}{5}$	$\frac{3}{4}$
31	4	4	L
30	34	34	L
29	23	3	2
28	5	1	15
27	L	L	L
26	4	4	L
25	234	34	2
24	3	36	6
23	5	L	5
22	4	4	L
21	34	34	L
20	23	3	2
19	5	1	15
18	L	L	L
17	4	4	L
16	234	34	2
15	3	36	6

14	5	L	5
13	4	4	L
12	34	34	L
11	23	3	2
10	5	1	15
9	L	L	L
8	4	4	L
7	234	346	26
6	3	36	6
5	5	L	5
4	4	4	L
3	3	3	L
2	2	1	12
1	L	1	1
0	L	L	L
−1	W	W	W
−2	W	W	W
−3	W	W	W
−4	W	W	W
−5	W	W	W

FIGURE **74** The result: the strategy grid for totals of 31 or less.

First Move	Face Showing	Total Remaining	Type of Position
1	1/6	30	W (play 3 or 4)
2	2/5	29	W (play 3)
3	3/4	28	W (play 1 or 5)
4	3/4	27	L
5	2/5	26	W (play 4)
6	1/6	25	W (play 2, 3, or 4)

"But there's a short cut to save doing all that. Just find row 31 and look in *all three columns* for the possible moves. See, *you get to choose* which face is on top for the first move, it isn't determined by the existing position. The moves in row 31, in fact, are 4 4 L. The only winning move among them is 4. That's consistent with the table I've just shown you, too.

"When *we* played, the total *was* 31, and I made the first move. Of course, being a perfect logician, I instantly worked out everything I've shown you so far, strategy grid included, so I *knew* I had to play 4. That left 27 as the remaining total. You replied with 5, taking the remaining total to 22. From the grid, remaining total 22 and face 2/5 is a winning move, and the strategy is to play 4. As I did, of course. You played 2 and I was faced with a remaining total of 16 and face 2/5. Now either 3 or 4 is a winning move: I played 3. You played 6, leading to a remaining total of 7 and face 1/6. Either 2, 3, or 4 is a winning move, and again I played 3. You *couldn't* play 4, to win, so you chose 1. Then I played my final winning move, another 3, and you were wiped out."

"Cunning. But for a big total you'd need to remember a rather long list!"

"Not really. The results from 17 onward just repeat what happens from 8 onward. The pattern of winning and losing moves repeats every nine rows, after the eighth. So the strategy for a total of 31 is the same as that for $31 - 9 = 22$ or $22 - 9 = 13$, and so on. For that matter, the strategy for 1012 is the same as that for $1012 - 9 \times (111) = 13$. All you need to remember is the first 17 rows of the grid!"

"Amazing," said Grimes, who was having a hard job remembering the first row, let alone the first 17.

"A mathematician would say that the pattern is periodic, and is determined by the total modulo 9," said Bumps. "That is, the remainder on division by 9. Except that you want to look in the range 8–16 if the number itself is bigger than 8. For example, the strategy for a total of 10 is *different* from that for 1, but that for 19 is the same as that for 10. The first few rows break the pattern because of end effects.

"So what you do is this:

- If the initial total is 8 or less, leave it alone.
- Otherwise, form its *digital root*, by repeatedly adding digits until you get a single number. (This is the same as its remainder on division by 9, except that 9 is used instead of 0.)
- If the digital root is 8 or 9 leave it alone.
- Otherwise, add 9.

Now you know which row to use to determine the strategy.

"For instance, let's take a starting total of 1012. It's bigger than 8, naturally. The digital root is $1 + 0 + 1 + 2 = 4$. That's not 8 or 9, so you add 9 to get 13. So the winning strategy for a total of 1012 is given in row 13 of the grid, as I already told you."

"Crikey!"

"See how powerful recursion is? It's solved the whole problem, for any initial total. Unlike the chocolate game, the perfect strategy is much harder to grasp as one unit. I don't think you could guess it in advance."

"You're right about that, Bumps."

"Indeed I am. Now, to test your understanding of my method, I want you to design a suitable set of cards, and then work out the winning strategy, for the same game—but played with a tetrahedral die, numbered 1–4. This time, it's the *downward* face that gets added to the total. Because there isn't a top face."

PROBLEM ❹

Grimes, to his surprise, managed it. How about you? I'll give you a clue: for a tetrahedral die, the pattern repeats with period 10. Which raises another interesting question for you to think about:

PROBLEM ❺

For games of this kind, where some object with numbered faces is repeatedly turned to change a total, does the strategy pattern always repeat? If the answer is yes, then there always exists a strategy that can be specified using a finite amount of paper.

I have in my possession a dodecahedral die, used for playing Dungeons and Dragons, numbered somewhat arbitrarily as in Figure 75. I wouldn't suggest using strategy cards, but it's not hard to program a personal computer to carry out the analysis. By the general result above, the strategy grid eventually becomes periodic. It turns out that the period is 26. The periodicity sets in from lines 1011 onward, which repeat line 985 onward.

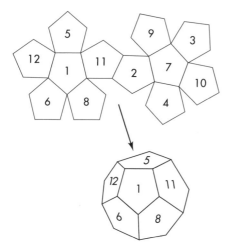

FIGURE **75**   A dodecahedral die. The corresponding grid has 12 columns.
The strategy repeats with period 26 from line 1011.

ANSWERS

1. *A.* With a total of 31 to aim at, and 6 faces on the die, the total
   number of possible positions is at most $32 \times 6 = 192$ (32 because 0
   is the starting total and 31 is legal). Therefore Rule 1 holds. Because
   the total increases at each stage, the game takes at most 32 moves:
   actually, it has to take less because you can't keep playing 1. The
   sequence of moves 1 2 1 2 . . . 1 2 1, with 21 moves, is the longest
   legal game. Anyway, Rule 2 holds. Rule 3 is clear, and so is Rule 4.
   B. Chess *can* end in a draw, so Rule 3 does not hold. Neither does
   Rule 4 (which is why chess puzzles tell you who is to move next).
   However, if we redefine "position" to be a pair (P, Q) where P
   represents the layout of the pieces on the board and Q specifies who
   is to move, we can satisfy Rule 4. Rule 1 is obeyed: with 32 pieces to
   place on 64 squares or off the board, there are at most 32^{65} positions, which is about 6.835×10^{97}. Rule 2 *can* be disobeyed: in
   principle a game can continue for an infinite number of moves (just
   move the two kings back and forth between two squares, once some
   room has been cleared to move them). However, one of the rules of
   chess says that either player can—but is not obliged to—claim a
   draw if the same position occurs three times. On the assumption

that a draw *is* always claimed under this law, the longest possible game takes at most $334^{12600} \sim 10^{104.6}$ moves.

2. The losing positions in Moldy Chocolate are the "square" shapes 1×1, 2×2, . . . , $n \times n$. The winning positions are all the others, the "rectangular" shapes. In fact, you can see this directly. Any rectangle can be cut once to leave a square (from any winning position we can produce a losing one); but whatever you do to a square it ends up rectangular (every move from a losing position produces a winning one).

   So Grimes starts in a winning position. His first move should be to break off a 4×6 piece, leaving 6×6. Thereafter, whatever Bumps does, Grimes can again leave her a square bar of chocolate, forcing her to end up with the moldy piece.

3. I learned the strategy for Divide and Conquer from Keith Austin of Sheffield University. Positions are given by two non-zero numbers (m, n), corresponding to m counters in one box and n in the other. The winning positions are those where at least one of m, n is even. The losing positions are those for which m and n are both odd.

   Let's check this. From a winning position, with at least one even number, play as follows: divide that even number between the two boxes so that each box contains an odd number of counters. For example, if n is even, divide it up into 1 and $n - 1$, both odd. Throw away the contents of the other box. This produces a losing position with both numbers odd.

   Given a losing position, with both numbers odd, then whichever number gets divided up will be odd. When an odd number is divided into two parts, one must be odd and the other even (odd + odd = even). Therefore the opponent is presented with a winning position again.

4. A possible set of cards for a tetrahedral die is shown in Figure 76 and the resulting strategy in Figure 77. Note that the pattern repeats from line 10 as shown by the arrow.

5. Does it always repeat? For this kind of game, where we number the faces of a polyhedron and roll it to produce a running total, the answer is yes. Suppose the largest number on any face is n. If a block of n consecutive rows in the strategy table is repeated, thereafter the entire table becomes periodic. (The strategy card has n rows to look at and so it "forgets" anything more than n moves

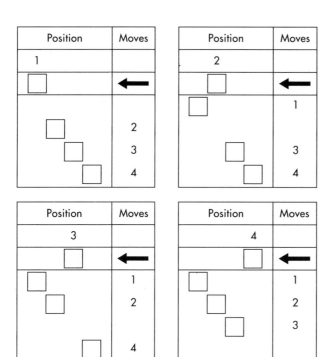

FIGURE 76 Strategy cards for rolling a tetrahedral die.

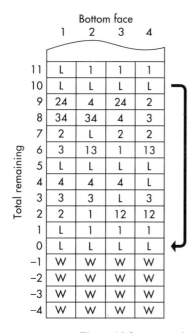

FIGURE 77 The grid for a tetrahedral die repeats every 10 rows.

back.) The number of such blocks is finite (but rather large) so a repetition is inevitable.

FURTHER READING

Berlekamp, R., J. H. Conway, and R. K. Guy. *Winning Ways.* New York and London: Academic Press, 1982.

Stewart, Ian. The number of possible games of chess. *Journal of Recreational Mathematics* 4 no. 1 (1971): 50.

———. *Game, Set, and Math.* Oxford and Cambridge, Mass.: Basil Blackwell, 1989; Harmondsworth, England: Penguin Books, 1991.

Thomas, L. C. *Games, Theory and Applications.* Chichester, England: Ellis Horwood, 1984.

CHAPTER 12

The Thermodynamics of Curlicues

"**F**our point five seven degrees right, darling . . . great . . . left hand down a bit . . . super, lovely, fantastic . . . eleven point six two degrees left, keep it up, it's going really well!"

Jacob Staff, the newly promoted borough surveyor of the market town of Upward-le-Mobile, clapped his hands against his head in anguish. He'd *told* the city architect not to hire Sandy Warthog to design the markings for the hypermarket car park. What it needed was a nice, simple grid of straight lines. But Warthog, whose main claim to fame seemed to be the creation of life-sized statues of trash cans—made by sticking a real trash can on a concrete pedestal—would have none of it. Warthog's critics considered his work to be rubbish, but the city architect had risen from the ranks of the Garbage Collection Service and thought Warthog was a genius.

Mind you, the design taking shape in yellow thermoplastic on the surface of the car park was remarkable: a complex system of spirals and curlicues (Figure 78), zigzagging this way and that, but not exactly at random. Jacob remembered arguing with Warthog about the design. "Straight lines, Sandy, that's what we need. Think about it logically. Nothing fancy." To which Warthog had replied, "Oh, but straight lines are so — well, so *linear*." And the city architect had agreed with Warthog — Jacob believed in the hope of winning an architectural design prize, which, he felt, always went to designs that were intellectually pretentious but looked pretty, rather than to anything functional. "I want to design something more *superficial*," Warthog had concluded.

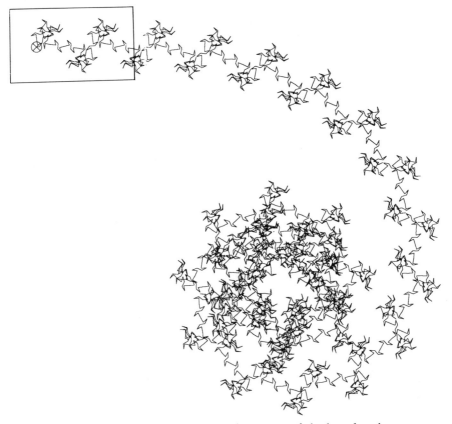

FIGURE **78**   Warthog's curve. For an enlargement of the boxed region, see Figure 86.

Jacob had always considered everything Warthog did as being superficial, and said so, but Warthog riposted by (a) accusing Jacob of being an ignorant slug and (b) defining "superficial" as "having dimension greater than one." "How," asked Jacob, "can a curve have dimension greater than one?" Instead of an answer, he was told that thanks to the imaginative research of Michel Mendès-France, not only could a curve have dimension bigger than one, it could also have entropy, temperature, volume, and pressure.

Mendès-France defined these quantities using a fruitful analogy with thermodynamics. We'll start with dimension, which is the simplest. In this age of fractals the idea of dimensions that are not integers causes few raised eyebrows; and it is reasonable that a sufficiently wiggly curve should be considered as having a larger dimension than a relatively linear one. Together with Michel Dekking, Mendès-France found a sensible way to achieve this, shown in Figure 79. Their concept is *not* the same as Hausdorff-Besicovitch dimension, commonly used to study fractals.

As Warthog explained to Jacob, a curve is said to be *linear* if its dimension is 1, and *superficial* if its dimension is greater than 1. For example, a logarithmic spiral (Figure 80 *A*) is linear, whereas an archimedean spiral (Figure 80 *B*) is superficial. Dekking and Mendès-France found an endless supply of superficial curves, constructed as infinite polygons. To explain their method, we first note that every real number x determines a unique direction in the plane, forming an angle $2\pi x$

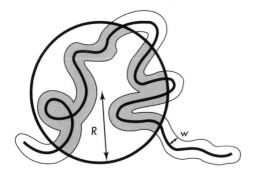

FIGURE **79**   The dimension of a curve according to Mendès-France. Surround the curve by a region of width w, and let A be the area of this region that lies inside a circle of radius R. The dimension is the limit of log A/log R as $R \to \infty$ and $w \to 0$, provided this exists.

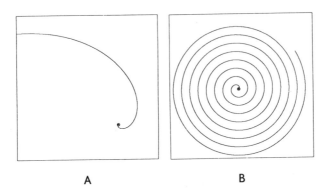

A   B

FIGURE **80**   *A.* The logarithmic spiral, whose equation in polar coordinates is $r = e^{\theta}$, has dimension 1. *B.* The archimedean spiral, whose equation in polar coordinates is $r = \theta$, has dimension greater than 1.

radians ($360x$ degrees) with the horizontal. The numbers x between 0 and 1 already provide all possible directions: moreover, adding an integer to x yields the same direction. That is, the direction depends only on the part of x that occurs after the decimal point, known as its fractional part $\{x\}$ or as x (mod 1).

"Now, Jake, old bean," said Warthog affably. "Choose your favorite sequence x_0, x_1, x_2,. . . . For example, you might let

$$x_n = \sin(\sqrt{n})$$

so that the sequence goes

$$0, 0.0174, 0.0246, 0.0302, \ldots$$

and so on. Then you construct a polygonal curve $\Gamma(x)$ as follows.

"From a chosen starting point, draw a line of fixed length in the direction corresponding to x_0. From there, draw one of the same length in the direction corresponding to x_1. From there, draw one of the same length in the direction corresponding to x_2, and so on, running through the entire infinite sequence."

With the above sequence this yields successive directions inclined to the horizontal at angles

$$0 \times 360 = 0 \text{ degrees}$$
$$0.0174 \times 360 = 6.264 \text{ degrees}$$

$$0.0246 \times 360 = 8.856 \text{ degrees}$$
$$0.0302 \times 360 = 10.872 \text{ degrees}$$

and so on. The start of the polygonal curve is shown in Figure 81 A, and a much longer segment—on a smaller scale—is shown in Figure 81 B. It is clear that this particular curve wanders off to the right, drawing larger and larger double loops. The loops grow slowly and stay separate from each other, so the curve is linear rather than superficial.

Each sequence $x = (x_n)$ yields a different curve $\Gamma(x)$: examples are shown in Figure 82. It was a curve of this kind that Sandy Warthog was drawing in the hypermarket car park.

PROBLEM ❶

There are no problems in this chapter—the heading is a mirage—but if you've got a personal computer you can emulate Warthog and draw curlicues for yourself. (I suggest you begin on paper and progress to car parks only with official permission.) A suitable algorithm is shown in the box entitled "Algorithm for Drawing $\Gamma(x)$" in pseudo-code (p. 185): you can implement it by hand (lots of work) or by computer, and choose your own sequence.

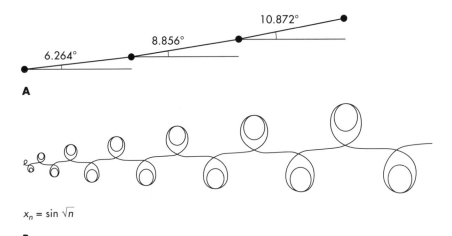

10.872°

8.856°

6.264°

A

$x_n = \sin \sqrt{n}$

B

FIGURE 81   *A.* The first three steps in the construction of $\Gamma(\sin \sqrt{n})$. *B.* Several hundred steps in the construction of $\Gamma(\sin \sqrt{n})$.

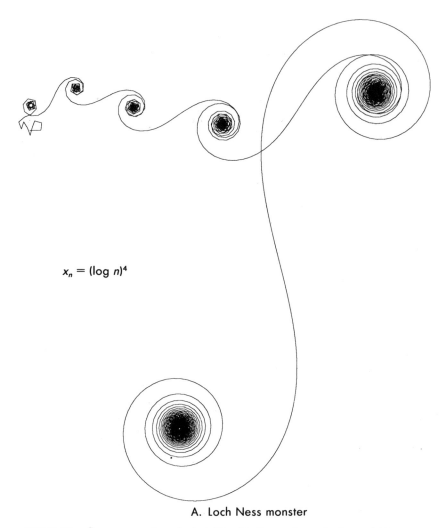

$$x_n = (\log n)^4$$

A. Loch Ness monster

FIGURE **82**   Some examples of what Mendès-France's method can achieve. *A*. The Loch Ness Monster, $\gamma((\log n)^4)$; *B*. Arches, $\gamma(n^{3/2})$; *C*. Bull, $\Gamma(n^3/1013)$; *D*. Bull ring, $\gamma(n^3/1002)$. (*Continued on pp. 183–84*)

Dekking and Mendès-France have characterized those sequences x that yield superficial curves. Precisely, they proved that the sequence x is "equidistributed modulo 1" if and only if all curves $\Gamma(mx)$ are superficial for all positive integers m. The sequence mx is the one whose terms

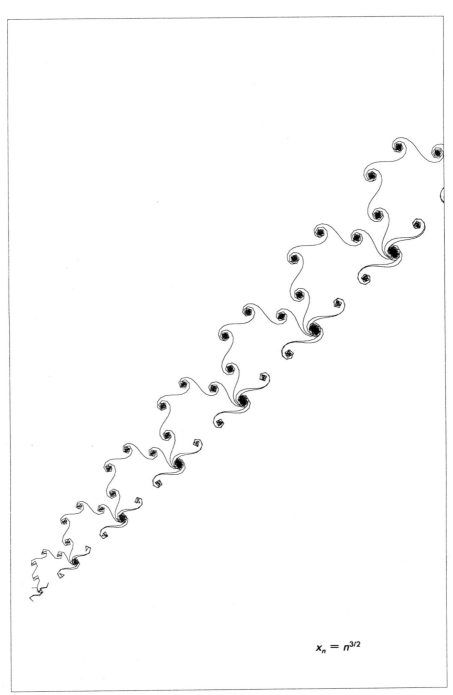

$$x_n = n^{3/2}$$

B. Arches

FIGURE **82** (*Continued*)

$x_n = n^3/1013$

C. Bull

$x_n = n^3/1002$

D. Bullring

FIGURE 82   (*Continued*)

ALGORITHM FOR DRAWING $\Gamma(x)$

I assume that x_n is specified by a formula or a separate sub-routine. Choose a starting point (sx, sy) and a scale factor sf, depending on your graphics system. The command *plot* draws a point with specified coordinates and *line* joins two points by a line. The algorithm is:

```
plot (sx, sy)
x = sx: y = sy
repeat for n = 0,1,2, . . . , finishing value:
    x1 = x + s cos(2πxₙ): y1 = y + s sin(2πxₙ)
    line (x, y) to (x1, y1)
    x = x1: y = y1
end repeat
```

are mx_0, mx_1 mx_2, . . . and so on. A sequence is equidistributed modulo 1 if the corresponding directions are equally distributed around the circle; that is, the probability that a randomly chosen term x_n in the sequence points in any given direction is the same for all directions. This is an important number-theoretic property of sequences, so their work links number theory to very curious geometric ideas. You can use it either way round: knowing that the sequence is equidistributed proves that the curve is superficial; knowing that the curve is superficial proves that the sequence is equidistributed. (WARNING: Usually *neither* is easy to establish in general!)

The thermodynamics of curves $\Gamma(x)$ is a little bit complicated, so it has been relegated to a box (see pp. 186–88). You don't have to read it. It is, in particular, possible to apply the ideas in the box to the curves $\Gamma(x)$ formed from sequences $x = (x_n)$, as described above, and investigate the thermodynamics of sequences.

Jacob Staff had a rather desperate idea. *A curve of zero temperature must be straight* . . . "Hey, Warthog! I'll tell you why you should use straight lines! They're real *cool* curves, man!" Unfortunately Warthog was unimpressed by outdated slang and responded instead with a challenge. "You said I ought to be logical about this project, Jake. OK, so *you*

THERMODYNAMICS OF A CURVE

Thermodynamics is the study of statistical properties of gases: it involves traditionally defined quantities such as temperature T, pressure P, volume V, and entropy S. Boyle's law for gases states that $PV = RT$ where R is a constant: it is valid at high temperatures T. We seek analogues of these ideas for curves rather than gases!

Let Γ be a curve of finite length. Consider an arbitrary straight line K in the plane: it meets Γ in a finite number n_K of points (unless, exceptionally, Γ contains a straight line segment that lies exactly along K). We say that K finds Γ in the state n_K. In thermodynamics the entropy of a system is defined to be

$$-[p_1 \log p_1 + p_2 \log p_2 + . . .]$$

where p_n is the probability that the system is found in state n. For curves we take p_n to be the probability that a randomly chosen line K finds Γ in the state n, that is, p_n is the probability that $n_K = n$. To make sense of this we require a suitable definition of probability on the space of all lines K. Results of Hugo Steinhaus lead to an explicit formula for the entropy of a curve, as follows. Let l be the length of Γ and let h be the length of its convex hull (the smallest convex curve that contains it: think of surrounding it with an elastic band and letting the band contract as far as it can). Then the entropy of the curve Γ is

$$S = \log\ (2l/h) + \frac{\beta}{e^\beta - 1}$$

where

$$\beta = \log[2l/2l - h)]$$

Whether the analogy makes sense or not, the quantity S is a well-defined property of the curve Γ, so we now take the formula as the *definition* of entropy. In traditional thermodynamics the temperature T is defined as the reciprocal of the quantity β, so we have

$$T = \{\log[2l/(2l - h)]\}^{-1}$$

Pursuing the analogy further, we define the volume V of the curve to be its length,

$$V = l$$

and the pressure P to be

$$P = h^{-1}$$

These definitions are geometrically appealing: the volume (3-dimensional measure) of a gas is replaced by the length (1-dimensional measure) of a curve; and the smaller the convex hull, the more compressed the curve must be to fit inside it, so the higher the pressure! From the definitions, the "gas law" relating these quantities is

$$2PV = (1 - e^{-1/T})^{-1}$$

which for high temperatures T yields

$$PV = \tfrac{1}{2}T$$

the analogue for curves, with a gas constant $R = \tfrac{1}{2}$, of Boyle's law!

For curves of infinite length, similar quantities are defined in terms of suitable limits of finite segments of the curves. In particular the entropy of such a curve Γ is defined to be

$$\lim_{r \to \infty} \frac{\log(2l_r/h_r)}{\log r}$$

where l_r is the length of a finite segment Γ_r of Γ and h_r is the length of its convex hull.

(Continued)

THERMODYNAMICS OF A CURVE (Continued)

The analogy bears interesting fruit. The entropy of a curve is always positive, and is zero if and only if the curve is a straight line. An algebraic curve, defined by a polynomial of degree d, has entropy at most $1 + \log d$. Thus entropy appears to be a natural measure of the complexity of the curve. This makes sense too: in information theory the entropy, as defined above, corresponds to the quantity of information (note that there are problems with sign conventions, and for some writers entropy is negative information). Thus the entropy, or complexity, of a curve can be interpreted loosely as the quantity of information needed to specify it.

It follows from the above equations that the temperature T of a curve is zero if and only if the curve is a straight line: at "absolute zero" temperatures, only straight lines exist. The hotter a curve becomes, the more wiggly it can be.

Of course, these analogies are very approximate, and any serious study must use the precise definitions!

be logical. I bet you can't work out what sequence I'm using to construct this curve. If I'm right, you stop behaving like a wimp. If I'm wrong, I'll rip it up again and draw a nice pattern of straight lines instead!"

"Come off it, Warthog! There's infinitely many possibilities! I wouldn't have a chance!"

"OK, I'll give you a hint. It's one of the sequences $\Gamma(a\,n^2)$ given by $x_n = a\,n^2$ for constant a. All you've got to do is work out the constant!" This particular family of sequences has deep connections with number theory (as we shall see), so the geometry of the associated curves ought to be interesting (as it is).

"But there's still an infinity of . . ." Jacob began. And stopped. Here was a serious chance—his *only* chance—to stop Warthog ruining the entire car park. A faint chance—the problem was still pretty difficult—but a chance, nonetheless. Jacob's mind went into overdrive. What makes Warthog's challenge hard is that different values of the constant produce a remarkable variety of curves (Figure 83). (For the moment ignore references to "renorm levels" in that figure; they are explained below.)

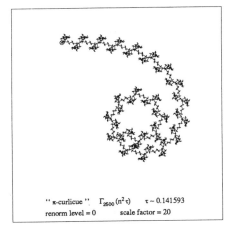

" π-curlicue " $\Gamma_{2500}(n^2\tau)$ $\tau \sim 0.141593$
renorm level = 0   scale factor = 20

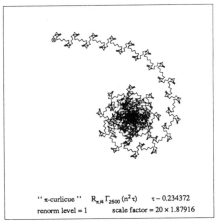

" π-curlicue " $R_{\pi/4}\,\Gamma_{2500}(n^2\tau)$ $\tau \sim 0.234372$
renorm level = 1   scale factor = 20 × 1.87916

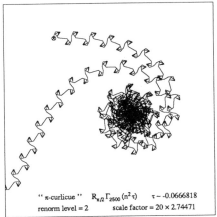

" π-curlicue " $R_{\pi/2}\,\Gamma_{2500}(n^2\tau)$ $\tau \sim -0.0666818$
renorm level = 2   scale factor = 20 × 2.74471

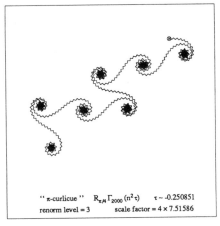

" π-curlicue " $R_{\pi/4}\,\Gamma_{2000}(n^2\tau)$ $\tau \sim -0.250851$
renorm level = 3   scale factor = 4 × 7.51586

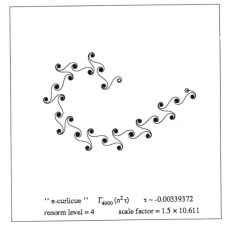

" π-curlicue " $\Gamma_{4000}(n^2\tau)$ $\tau \sim -0.00339372$
renorm level = 4   scale factor = 1.5 × 10.611

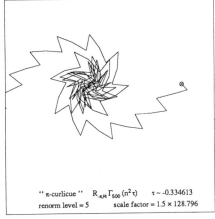

" π-curlicue " $R_{-\pi/4}\,\Gamma_{500}(n^2\tau)$ $\tau \sim -0.334613$
renorm level = 5   scale factor = 1.5 × 128.796

FIGURE **83**   Twelve curves of the form $\Gamma(an^2)$. (*Continued on next page*)

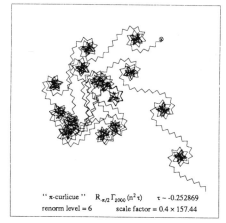

" π-curlicue " $R_{-\pi/2}\,\Gamma_{2000}\,(n^2\tau)$ $\tau \sim -0.252869$
renorm level = 6   scale factor = 0.4×157.44

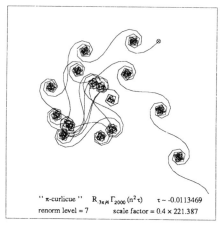

" π-curlicue " $R_{-3\pi/4}\,\Gamma_{2000}\,(n^2\tau)$ $\tau \sim -0.0113469$
renorm level = 7   scale factor = 0.4×221.387

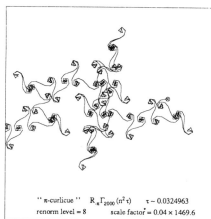

" π-curlicue " $R_{-\pi}\,\Gamma_{2000}\,(n^2\tau)$ $\tau \sim 0.0324963$
renorm level = 8   scale factor = 0.04×1469.6

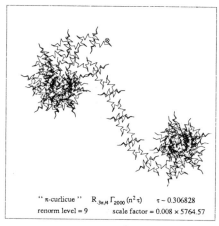

" π-curlicue " $R_{-3\pi/4}\,\Gamma_{2000}\,(n^2\tau)$ $\tau \sim 0.306828$
renorm level = 9   scale factor = 0.008×5764.57

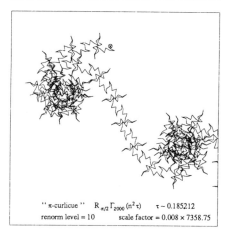

" π-curlicue " $R_{-\pi/2}\,\Gamma_{2000}\,(n^2\tau)$ $\tau \sim 0.185212$
renorm level = 10   scale factor = 0.008×7358.75

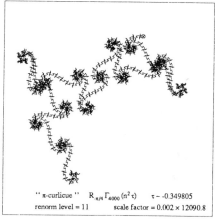

" π-curlicue " $R_{-\pi/4}\,\Gamma_{4000}\,(n^2\tau)$ $\tau \sim -0.349805$
renorm level = 11   scale factor = 0.002×12090.8

FIGURE 83   (*Continued*)

When faced with a complicated range of possibilities, the best approach is to start from simple cases and see if any general principles emerge. The simplest case of all is when a is the reciprocal $1/N$ of an integer N. The corresponding curves $\Gamma(n^2/N)$ have a very simple spiral structure (Figure 84), closely resembling the Cornu spiral, which is important in optics. This is not an accident: the process of drawing $\Gamma(n^2/N)$ can be viewed as a discrete approximation to a Cornu spiral. Other values of a produce more complex curves, but repeated spiral structures, or curlicues, are always prominent features.

The car park was a maze of curlicues. Might they provide a clue?

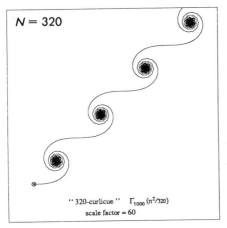

N = 320

"320-curlicue" $\Gamma_{1000}(n^2/320)$
scale factor = 60

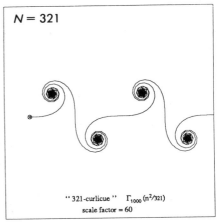

N = 321

"321-curlicue" $\Gamma_{1000}(n^2/321)$
scale factor = 60

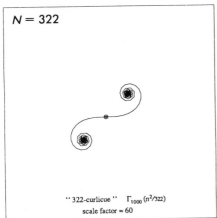

N = 322

"322-curlicue" $\Gamma_{1000}(n^2/322)$
scale factor = 60

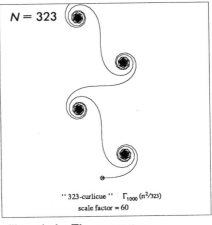

N = 323

"323-curlicue" $\Gamma_{1000}(n^2/323)$
scale factor = 60

FIGURE 84   Curves $\Gamma(n^2/N)$ form Cornu-like spirals. The symmetry depends on n (mod 4), and four representative cases are shown.

Jacob had to work out what causes the curlicues. Let's look at the special case a $= 1/N$, where N is a fairly large integer. Consider the nth phase difference, the angle between the directions of the nth and $(n + 1)$st segments in the polygon $\Gamma(n^2/N)$. In radian measure this is equal to $((n + 1)^2 - n^2)2\pi/N = (2n + 1)2\pi/N$. When n is small the phase difference is also small, so the segments point in almost the same direction, producing a gentle curve. When the phase difference becomes greater than $\pi/2$ (a right angle) the segments fold back on each other, filling in a blob. As n approaches $N/4$ each segment almost reverses its predecessor. As n increases beyond $N/4$ the curlicue begins to unwind again, and when n reaches $N/2$ it straightens out, after which it starts to form another curlicue. After N steps the pattern repeats. There are four cases, depending on the value of N (mod 4); that is, on whether N is of the form $4m$, $4m + 1$, $4m + 2$, or $4m + 3$: this is again illustrated in Figure 84.

The explanation of this phenomenon comes from the theory of Gauss sums. A "complete" Gauss sum, which is most simply expressed in complex notation, takes the form

$$\sum_{n=0}^{N} e^{2\pi i, n^2}/N \tag{1}$$

where the large sigma indicates summation and $i = \sqrt{-1}$. Its value is known to be

$$2(1 + (-i)^N) \frac{1 + i}{4} \sqrt{N} \tag{2}$$

We can think of the plane in which the curve $\Gamma(n^2/N)$ is drawn as being the complex plane, and then we find that the rth vertex of the curve is the partial Gauss sum obtained from (1) above by changing the upper limit of the range of summation from N to r. Suppose that N divides r exactly p times, with remainder q: then $r = pN + q$ and the Gauss sum to r terms is that to q terms, plus p complete Gauss sums. It can then be seen that (2) is responsible for the various periodicities and symmetries in Figure 84. The qualitative behavior depends only on N (mod 4) because $(-i)^N$ depends only on N (mod 4). In fact

$$(-i)^N = 1 \quad \text{if } N \neq 0 \ (\text{mod } 4)$$
$$= -i \quad \text{if } N \neq 1 \ (\text{mod } 4)$$
$$= -1 \quad \text{if } N \neq 2 \ (\text{mod } 4)$$
$$= i \quad \text{if } N \neq 3 \ (\text{mod } 4)$$

The next simplest case is when a is a rational number, say $a = p/q$ for coprime integers p, q. These rational approximations produce periodic spiral structures in the curve, much as $a = 1/N$ did above. Figure 85 shows what happens for $p/7$, when $p = 1, 2, 3, 4, 5, 6$.

Next Jacob realized that he could use the structure of $\Gamma(\frac{p}{q} n^2)$ for rational a to explain some of the features observed for irrational values of a. Suppose that $\frac{p}{q}$ is a good rational approximation to a; then the shape of the beginning of the curve $\Gamma(an^2)$ will closely resemble that for $\Gamma(\frac{p}{q} n^2)$. Jacob could make use of this idea to guess what value of a Sandy Warthog was using. Which rational approximations can you see in Warthog's curve (Figure 78)?

To help, the beginning of that curve (the boxed part in Figure 78) is shown, enlarged, in Figure 86. The windings are rather complicated and tend to overlap, so I'll give you a couple of hints. There is an approximately repeating zigzag of length 7, corresponding to a rational approximation of the form $p/7$ for some p. Comparing the shape of the zigzag

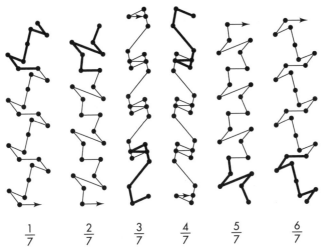

$$\frac{1}{7} \qquad \frac{2}{7} \qquad \frac{3}{7} \qquad \frac{4}{7} \qquad \frac{5}{7} \qquad \frac{6}{7}$$

FIGURE 85   The curves $\Gamma(pn^2/7)$ for $p = 1, 2, 3, 4, 5, 6$ all exhibit repeating segments of length 7, the first such segment being shown as a solid line. The curves for p and $7-p$ are identical except for a rotation of $180°$.

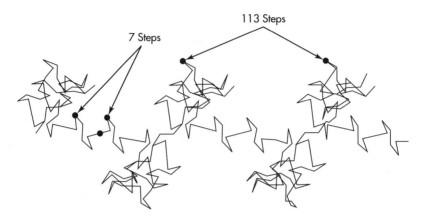

FIGURE **86**   Close-up of the boxed region of Figure 78, Warthog's curve, showing almost-repeating structures of lengths 7 and 113.

with Figure 85, we see that p must be congruent to 1 or 6 modulo 7, so the rational approximation is one of 1/7, 6/7, 8/7, 13/7, 20/7, 22/7, 27/7, 29/7 . . . Notice anything familiar in that list? To clinch the guess, Figure 86 also shows a larger, approximately repeating structure, whose length is the unusual integer 113. Do you know any interesting rationals of the form $p/113$?

Of course, it's π. The first good rational approximations to π are 22/7 and (far better) 355/113. So it looks as though Warthog's curve is $\Gamma(\pi n^2)$. The repeated zigzags reflect the approximation 22/7; the large minicurlicues come from the 355/113. In this way the number theory of rational approximations becomes related to the geometry of the corresponding curve.

"You're using the value a $= \pi$!" yelled Jacob in triumph.

Warthog's face fell. "How did you — "

"I win! Now, get rid of those silly curlicues and make me a nice linear grid like you promised!"

His face the perfect picture of suppressed fury, Warthog set to work.

The approximate periodicities that led Jacob to his solution can be investigated further using another idea from physics. If you look at the curve $\Gamma(\pi n^2)$ from a great distance you effectively suppress the minicurlicues corresponding to the rational approximation 22/7 and see only the spiral given by 355/113. It can be proved that long initial segments of curves $\Gamma(an^2)$ are well approximated by shorter initial segments of curves $\Gamma(bn^2)$, where b is a different constant. In fact, we can assume that a lies between $-\frac{1}{2}$ and $\frac{1}{2}$, because adding an integer to a or subtracting an integer from a has no effect on $\Gamma(an^2)$. If we take

$$b = \left\{ \frac{1}{2a} \right\} - \frac{1}{2a}$$

where $\{x\}$ is the fractional part of x, then we can approximate the first N segments of $\Gamma(an^2)$ by the first $2aN$ segments of $\Gamma(bn^2)$. The picture for $\Gamma(bn^2)$ must be scaled by a suitable amount and turned through a right angle. This result was proved by the physicists Michael Berry and J. Goldberg in 1988; it has applications to quantum mechanics and optical diffraction. The process is called renormalization, and it originated in quantum theory—though it should be added that in 1914 the number theorists Godfrey Hardy and John Edensor Littlewood applied a weaker version of the same idea to estimate the values of partial Gauss sums.

Figure 83 actually shows successive stages in the renormalization of $\Gamma(\pi n^2)$. The resemblances between successive stages are sometimes obvious—for example, the first three pictures have very similar shapes. When the resemblances are less clear this happens because the scale of the drawing has changed: this is shown by the first number after the words "scale factor." The second number indicates the natural scaling involved in the renormalization process. For example, the fifth and sixth pictures use the same drawing scale (1.5) but combine it with different renormalization scalings (10.611 and 128.796). Pictures in which the first scaling number is the same have directly comparable shapes.

Two days later the curve $\Gamma(\pi n^2)$ had been erased from the car park, and Warthog was about to lay down the promised linear grid. The city architect glared at Jacob. "I still think we should have left the decision up to Warthog! Brilliant man . . . superb imagination—"

Jacob interrupted him. The new pattern—Figure 87—didn't look quite right. "Hey! Warthog!" he yelled. "What are you up to *now*? I thought you were going to produce a grid of straight lines!"

Sandy Warthog looked up from his work. "It *is* a grid of straight lines," he said.

"Yes, but they ought to form car-sized rectangles!"

FIGURE **87**   Warthog's revenge, $\Gamma(n^7/1050)$.

Warthog shook his head sadly. "A bit late to tell me *that*, Jake. Not after I spent half the night coming up with this design for you. I thought you'd be *pleased*! There's just no gratitude in this world."

The city architect beamed from ear to ear.

Jacob Staff began to sob quietly.

ANSWERS

The answers are a mirage, too.

FURTHER READING

Berry, M. V., and J. Goldberg. Renormalization of curlicues. *Nonlinearity* 1 (1988): 1–26.

Dekking, M., and M. Mendès-France. Uniform distribution modulo one: a geometrical viewpoint. *Journal für die Reine und Angewandte Mathematik* 329 (1981): 143–53.

Dekking, M., M. Mendès-France, and Alfred J. van der Poorten. Folds! *Mathematical Intelligencer* 4 (1982): 130–38, 173–81, 190–95.

Deshouillers, Jean-Marc. Geometric aspects of Weyl sums. In *Elementary and Analytic Theory of Numbers*. Warsaw: Banach Centre Publications, 1985.

Hardy, G. H., and J. E. Littlewood. *Acta Mathematica* 37 (1914): 192–239.

Moore, Ross R., and Alfred J. van der Poorten. On the thermodynamics of curves and other curlicues. McQuarie University Mathematics Reports 89-0031: April 1989. Forthcoming in *Proceedings of Conference on Geometry and Physics*. Canberra, 1989.

CHAPTER 13

•

The

•

Group-Theorist

•

# of Notre Dame

•

—————

•

Quasimodulo, the mathematical hunchback of Notre Dame, was perched on an oak beam midway between the largest bell in the cathedral and its neighbor. He held a slim form tenderly in his muscular arms. The form opened one eye, blinked, gasped, and began to struggle wildly.

"Esmeralda?" asked Quasimodulo hopefully.

"Good heavens, no!" said the slim form. "She's my maid. I'm afraid, sir, you have made a mistake!"

I'm getting too old for this, thought Quasimodulo. "Zen 'oo are you?" he said.

"I can't tell you, we haven't been introduced," said the slim form's owner primly. Quasimodulo relaxed his hold slightly. She glanced down at the sheer drop, and grasped his chest tightly. "On the other hand, I was never one for formalities. I am Jane Porter, the daughter of Professor Archimedes Q. Porter! Um — you don't by any chance, usually swing through trees instead of belfries, do you? I was rather keen to find someone who swung through trees . . ." She looked at him inquiringly. "You wouldn't be a well-muscled bronzed young god of a man called Tarzan, would you? Does the name Greystoke mean anything to you? I mean, you *do* dangle from ropes . . . you haven't just swung in from Africa by any . . ." she tailed off lamely, eyeing his hunch and his ragged clothes. Tarzan was never renowned as a master of disguise. Of course, the creature *might* be one of his apes . . .

"Zut alors," said Quasimodulo. "I fear we 'ave both made an unfortunate mistook." He carried her to a stone ledge, well clear of the precipitous drop to the stone flags beneath. "Please accept ze apologies, Mlle. Jane. Zis eyesight of mine, she gets badder every day." Jane stared at him and he found himself unable to meet her gaze. To ease his embarrassment he began to swing from bell to bell, ringing them more and more loudly, bong-bong-dong-ding-*bong-dong*-BONGGG!!!!

"Anyone can see you're no campanologist," said Jane haughtily.

"Non, ma'm'selle, I am a Catholic."

"I mean, you're not a bell-ringer."

Quasimodulo looked at her, then at the bells, then back at Jane. "What do you sink *zose* are, ma'm'selle? Ze tea-cosies?"

"Oh, I agree that you're ringing bells, but that doesn't make you a bell-ringer. Not in the manner that I intended the phrase. In *England"* —she paused to give the word emphasis—"we take our bell-ringing very seriously. I was listening carefully, and I noticed that you rang the tenor bell at least three times and didn't ring the treble at all!"

"Is zat bad?" asked Quasimodulo, puzzled.

"You broke the cardinal rule of bell-ringing, sir!" After which Jane had to explain at some length what she meant.

Bell-ringing is a popular hobby in England. Teams of ringers gather regularly at churches to ring changes. This means that they must ring a given set of bells in different orders, until all possible orders have been rung exactly once.

"A simple problem in ze mathematics," muttered Quasimodulo.

"Good heavens, the creature is *educated*!" cried Jane in surprise.

"Mine is a simple tale," said Quasimodulo. "I was once professeur des mathématiques at ze Université de Paris. I work on ze vibrations of solid objects: ze bells, zey were a passion of mine. Oh, ze bells, ze bells! Zere is such a wild beauty in ze vibrational spectrum of a bell! Ze Bessel functions, ze eigenvalues of ze Laplacian! But wiz my increasing deformity — which I blame upon carrying around large treatises on ze wave equation — my 'unch make it impossible to put chalk to ze blackboard, so I am forced to accept early retirement. I wander ze streets, a starving, miserable wreck, but all ze time I hear ze bells of Notre Dame calling me, calling me . . . One night when ze moon is full I climb ze outside of ze tower, and now I am so practiced at zat manoeuvre zat no one 'as been able to catch me."

"A sad tale," said Jane.

"Not completely. Ze life 'as its good points. At least I don't 'ave to grade examination papers any more. But zis idle bavardage puts my back up. You were talking about bells."

"Indeed I was," said Jane. "I am well versed in the mathematical art, thanks to my tutors in England."

"Zen ma'm'selle will not take it amiss if I declare zat ze problem is trivial? For if zere is n bells, zen ze number of distinct orderings is ze factorial of n, denoted

$$n! = n(n-1)(n-2). . .3 \cdot 2 \cdot 1$$

and zese can be rung in any sequence whatsoever. Thus zere is $(n!)!$ distinct ways to ring changes on n bells, a number zat grow très rapidement — "

PROBLEM ❶

According to Stirling's formula, $n!$ is approximately $\sqrt{2\pi n}\ n^n e^{-n}$. Find (to the nearest power of 10) the value of n for which the number of ways to ring a full set of changes on n bells is approximately one googolplex ($10^{10^{100}}$).

"Ah, but you didn't let me state all the rules," said Jane. "Denote the bells by the numbers 1, 2, 3, . . . , n, arranged in descending order of pitch. Bell 1 is called the treble, bell n the tenor. Each ordering of the bells, that is, each arrangement of the symbols 1, 2, 3, . . . ,n, is called a change. The ordering down the scale, that is, 123. . .n, is said to be *in* rounds. Finally, a complete sequence of all possible changes (plus one, a repeat of the first) is called a peal. There are five rules for ringing a full peal:

1. The peal must begin and end in rounds.
2. In between it must go through all possible changes without repetition.
3. No bell may move more than one place between successive changes.
4. No bell may stay in the same place for more than two successive changes.
5. Each bell should move in an equally varied manner.

"Rule 1 is for musicality," Jane pointed out.

"Ringing down ze scale," said Quasimodulo. "I see. So ze sequence 123. . .*n* appears *twice*: once at ze beginning and once at ze end. But all ze other orderings—"

"*Changes*, Quasimodulo! Please use the proper terminology!"

"—*changes* occur exactly once."

"Precisely! Rule 2 is mathematically satisfying, but it also has a practical purpose: to produce maximum variety."

"I suppose Rule 3 is for mechanical reasons," said Quasimodulo. "A bell 'as très grand momentum and ze period between successive rings cannot easily be shortened or lengthened." He paused. "And don't I know it," he added.

"Correct! Rule 4 is for variety, and Rule 5—which I admit is a trifle vague—for symmetry. I should add that Rules 4 and 5 are sometimes relaxed, for artistic reasons.

"The different number of bells used in practice are given special names: 3 is singles, 4 minimus, 5 doubles, 6 minor, 7 triples, 8 major, 9 caters, 10 royal, 11 cinques, and 12 maximus. For example, the method called Glasgow Surprise Major must refer, by virtue of the final word in its name, to changes rung on eight bells."

The beginning of Glasgow Surprise Major is shown in Figure 88. The lines show the movements of each bell: note that no bell ever moves more than one place to left or right in the sequence. On the other hand, no bell stays in the same place for more than three changes. But some bells do stay in place that long—for example, bell 6 stays in position 6 for the first three changes. The full peal contains $8! + 1 = 40,321$

Change   Movement of bells

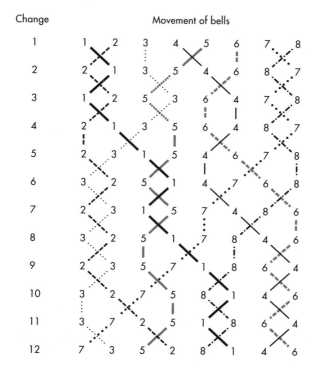

FIGURE **88**   The start of Glasgow Surprise Major.

changes, so it would be inconvenient to list it here, but I hope it conveys the general idea.

"Parfaitement!" exclaimed Quasimodulo. "Already I 'ave prove a theorem! Wiz three bells zere is only two possible peals!"

"If it is a theorem then it must have a proof."

"It 'as. It is easy to see zat each change must just swap two bells: you cannot move all three and at ze same time move each only one place. Ze middle bell cannot be leave in place—zat would mean ze two end bells interchange, each moving two places, contrary to Rule 3. So all changes must either swap ze first two bells, or swap ze last two. Ze same swap cannot be perform twice in succession, because zat would just repeat a change. Therefore ze two swaps must be performed alternately. If we start by swapping 12 zen we get

$$123\ 213\ 231\ 321\ 312\ 132\ 123$$

and if we start by swapping 23 we get

$$123 \ 132 \ 312 \ 321 \ 231 \ 213 \ 123$$

which — I now notice — is just ze first sequence of changes in reverse."

"Excellent! The first sequence you listed is called quick six and the second is called slow six."

"Funny. If one she is ze reverse of ze other, zey should each take ze same time!"

"Probably psychological: one *sounds* slower." Jane looked thoughtful. "You know, your proof that there are only two different peals with three bells suggests to me that there must be some mathematical structure behind all this."

"Well . . . I could represent my proof wiz a *graph*. Suppose I draw six blobs, each corresponding to one possible sequence of ze symbols 1, 2, 3. Zose are ze six changes. Zen I join changes by a dotted line if I can get from one to ze other by swapping ze first two bells, and by a solid line if I can get from one to ze other by swapping ze second two bells. Ze problem of ringing all six changes becomes a problem in graph theory: find a 'amiltonian circuit, zat is, a closed loop which passes through each blob once and once only. 'Ere it is easy because ze entire graph forms one circuit [Figure 89]. Ze two solutions traverse ze circuit either clockwise or counterclockwise."

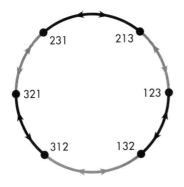

FIGURE **89**   Graph of possible moves from one change to the next with three bells. Dotted line = "swap first two bells," solid line = "swap last two bells."

"Mmmm . . . But finding Hamiltonian circuits is an unsolved problem in general. Anyway, I think there's *more* structure than that!"

"Per'aps she is ze group theory."

"Yes! Of course! Quasimodulo, you're a genius! How did you work
that out?"

"Just guess. It always pay to back a 'unch."

Group theory is about sets of transformations which, when performed in succession, yield another transformation in the same set.
(There's a more abstract definition but this one will suffice here.) You'll
see what I mean by this in a minute. The key to the mathematics of
bell-ringing is to focus not on the particular sequence of bells that occur
in any given change, but in the transformation that rearranges them.

Define two transformations F and L as follows:

$$F = \text{"interchange the } first \text{ two bells"}$$

$$L = \text{"interchange the } last \text{ two bells"}$$

For example, if we have the change

$$312$$

and apply F then we get

$$132$$

whereas applying L creates

$$321$$

The transformations F and L are illustrated in Figure 90*A*. Also, let I
denote the *identity* transformation:

$$I = \text{"leave all bells unchanged"}$$

Then the two peals "quick six" and "slow six" take on a very regular
form, shown in the box entitled "Quick and Slow Six in Terms of
Transformations" on p. 208. This box needs some explanation. Each
successive change is numbered, and the sequence of bells is listed. The

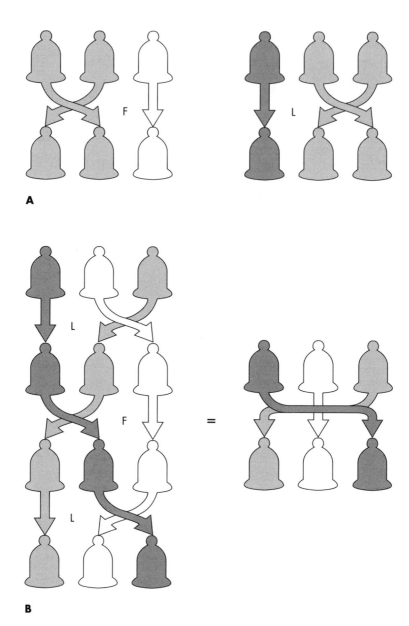

FIGURE **90**   *A.* The permutations F and L. *B.* The composite permutation LFL. *C.* Proof that LFLFLF = I.

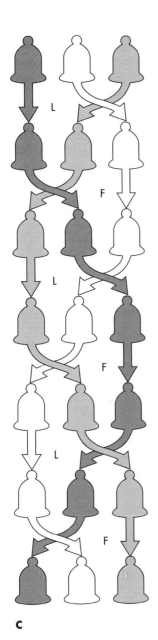

C

FIGURE **90**   (*Continued*)

QUICK AND SLOW SIX IN TERMS OF TRANSFORMATIONS

Quick Six

Change	123	Move	Permutation
1	123	I	I
2	213	F	F
3	231	L	LF
4	321	F	FLF
5	312	L	LFLF
6	132	F	FLFLF
7	123	L	LFLFLF

Slow Six

Change	123	Move	Permutation
1	123	I	I
2	132	L	L
3	312	F	FL
4	321	L	LFL
5	231	F	FLFL
6	213	L	LFLFL
7	123	F	FLFLFL

transformation used to obtain that sequence from the previous change is given in the "move" column — the entry I is purely conventional since there isn't a previous change — and the combined effect of all operators used is written in the final column under the heading "permutation." By this we mean the transformation that produces the given change from the original sequence 123. Each permutation is written as a string of F's and L's. These strings must be read from *right to left*: for example LFL means "first perform L, then F, then L again": see Figure 90B.

For definiteness, let's concentrate on slow six. Similar remarks apply to quick six by interchanging L and F.

The seventh change is the same as the first — that is, the string FLFLFL produces the same effect as the transformation I. We write this symbolically as

$$FLFLFL = I$$

or

$$(FL)^3 = I$$

which is illustrated in Figure 90*C*.

Indeed, *any* string composed of Ls and Fs must lead to precisely one of the six different possible changes, so any such string is equal to one of the six strings I, L, FL, LFL, FLFL, LFLFL. In particular, if we combine two of these strings together, we get a string which must equal one of the six just listed. For example

$$(LFLFL)(FLFL)$$

is the same (omit brackets) as

$$LFLFLFLFL$$

But this (putting the brackets in differently) is

$$(LFLFLF)(LFL)$$

and LFLFLF $=$ I, so we obtain

$$(I)(LFL)$$

which equals

$$LFL$$

because I doesn't alter anything. That is,

$$(LFLFL)(FLFL) = LFL$$

What about (LFLFL)(LFL)? We can drop the brackets and write this as

$$\text{LFLFLLFL}$$

but now what?

To make further progress, observe that LL = I because L swaps the last two bells, and doing this twice restores everything to its original position. We can write this as $L^2 = I$. Similarly, $F^2 = FF = I$. So we have

$$\text{LFLFLLFL} = \text{LFLFIFL} = \text{LFLFFL} = \text{LFLIL} = \text{LFLL} = \text{LFI} = \text{LF}$$

In this way we can build a complete multiplication table for the six strings listed above: see the box below entitled "Multiplication Table for Strings of Ls and Fs."

PROBLEM ❷

In deriving the box below there is one useful trick that you may not spot. The transformation F must be one of the six listed, but which?

MULTIPLICATION TABLE FOR STRINGS OF Ls and Fs

1st string **2d string**	I	L	FL	LFL	FLFL	LFLFL
I	I	L	FL	LFL	FLFL	LFLFL
L	L	I	LFL	FL	LFLFL	FLFL
FL	FL	LFLFL	FLFL	L	I	LFL
LFL	LFL	FLFL	LFLFL	I	L	FL
FLFL	FLFL	LFL	I	LFLFL	FL	L
LFLFL	LFLFL	FL	L	FLFL	LFL	I

Note: The combination is worked out in the order (2d string)(1st string).

The box shows that *every* product of the six strings I, L, FL, LFL, FLFL, LFLFL is another one of those six strings. That is, the set of six strings forms a group. It is called the symmetric group S_3, and it consists of all six possible permutations of three symbols. Its structure — the multiplication table — is completely determined by four equations:

$$IX = X = XI \text{ for any string } X$$

$$L^2 = I$$

$$F^2 = I$$

$$(LF)^3 = I$$

PROBLEM ❸

Work out the group table for quick six in the same way.

Let's be more general. Recall that a permutation of n objects arranged in order is a transformation that rearranges them. The usual notation for a permutation is a bracket containing two rows of symbols:

$$\begin{pmatrix} \text{initial order} \\ \text{final order} \end{pmatrix}$$

For example

$$\begin{pmatrix} 1234567 \\ 2317564 \end{pmatrix}$$

means:

move the first symbol to the second place

move the second symbol to the third place

move the third symbol to the first place

move the fourth symbol to the seventh place

leave the fifth symbol in the fifth place

leave the sixth symbol in the sixth place

move the seventh symbol to the fourth place

The effect of this is shown in Figure 91. It is important to distinguish between a *particular* arrangement of the symbols 1,2,3,. . ., *n* and a permutation: a permutation is a transformation that can be applied to *any* sequence of the symbols to produce a new sequence. (This can be confusing because an arrangement is often called a permutation in ordinary language.) For example, the permutation in Figure 91, when applied to the sequence of letters ABCDEFG yields the sequence CABGEFD; when applied to the sequence of numbers 7654321 it yields 5761324, and so on.

Permutations can be combined by performing them in turn, as shown in Figure 92. The *n*! possible permutations on *n* symbols form a group known as the symmetric group of degree *n*, written S_n.

There is a more compact notation for permutations, known as cycle notation. For example, the cycle

$$(345)$$

means

move the third symbol to the fourth place

move the fourth symbol to the fifth place

move the fifth symbol to the third place

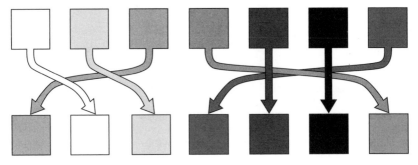

FIGURE **91**   Effect of a permutation.

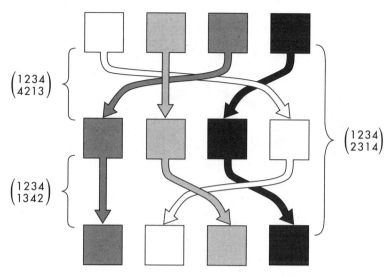

$$\begin{pmatrix} 1234 \\ 4213 \end{pmatrix}$$

$$\begin{pmatrix} 1234 \\ 2314 \end{pmatrix}$$

$$\begin{pmatrix} 1234 \\ 1342 \end{pmatrix}$$

FIGURE **92**   How to combine two permutations to get a third.

as in Figure 93*A*. Notice how the final move wraps around back to the start of the cycle, as in Figure 93*B*. Applied to eight bells in the sequence

$$12345678$$

the cycle (345) produces the new sequence

$$12534678$$

Applied to the sequence

$$35718243$$

it produces

$$35871243$$

and so on.

More generally, a product of cycles such as

$$(345)(26)$$

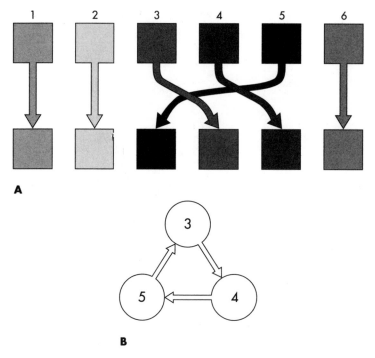

FIGURE 93   *A.* The cycle (345) as a permutation. *B.* How it wraps around.

means

> move the third symbol to the fourth place ⎫
> move the fourth symbol to the fifth place ⎬ cycle (345)
> move the fifth symbol to the third place ⎭
>
> move the second symbol to the sixth place ⎫
> move the sixth symbol to the second place ⎬ cycle (26)

and so on.

Which transformations can be used to move from one change to the next in bell-ringing? It is not hard to see that they must be formed by combining cycles of length 2, which interchange *adjacent* bells, such that no two of the cycles overlap. For example, on eight bells we might use (34), (23)(56) or (12)(45)(67), but not (13) or (2345)(see Figure 94).

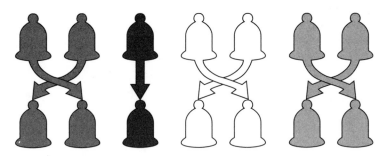

FIGURE **94**   Possible moves from one change to the next must interchange non-overlapping pairs of neighboring bells.

In this notation we can look at a more complicated peal, the minimus, rung on four bells, which is represented graphically in Figure 95. Here we use the transformations A = (23), B = (34), and C = (12)(34) in a regular pattern. The diagram breaks up into three loops, and A and C are used alternately around each loop. The irregular move B is used to switch from one loop to the next.

"'Ey!" yelled Quasimodulo. "I 'ave notice something! If we stop just before zat irregular move, and look only at ze first eight changes, zen ze permutations zat produce zem forms a smaller group in zeir own right! A *subgroup!*"

"You're right," said Jane. "Let's call it the hunting subgroup, and denote it by H."

"Why? Oh, I know, you English nobility is obsessed wiz 'unting."

"No, no! A bell is said to *hunt* if it moves steadily in one direction, one place at a time. And in H all three bells hunt in both directions in turn."

The hunting subgroup H plays a crucial role in the entire peal of minimus. Not only does it give the first eight changes—the *next* eight changes correspond to permutations that can be written as the product h(243) where h is a permutation in the hunting subgroup H. This set is called the *coset* H(243). Similarly, the final eight changes lie in the coset H(234), that is, they are obtained using permutations of the form h(234) where h is in H. In other words, the pattern in minimus corresponds to splitting the full group of all 24 permutations into three cosets corresponding to the subgroup H. In Figure 94 each loop of eight changes corresponds to a coset.

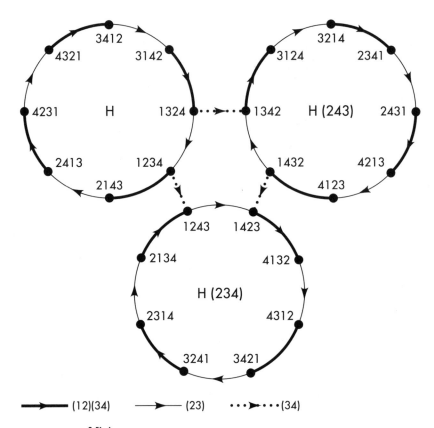

FIGURE **95**   Minimus.

"That was a brilliant observation, Quasimodulo!" Jane cried, having worked all this out. "I now see how to obtain Stedman's Doubles group-theoretically!"

Stedman's Doubles is a method used with five bells. The possible transformations from one change to the next are A = (23)(45), B = (12)(34), C = (12)(45), D = (12), E = (23), F = (34), and G = (45). The last four transformations are called singles by bell-ringers, for obvious reasons.

There are 5! = 120 changes on five bells. The method starts by applying C and A in turn, producing a subgroup containing only 6 changes. Stedman's Doubles splits the full group into 20 cosets each consisting of 6 permutations, and uses B to link them together into a

STEDMAN'S DOUBLES

Read down columns in turn. Horizontal lines indicate cosets. The final coset wraps around to the start.

Change	Move	Change	Move	Change	Move	Change	Move
12345	I						
21354	C						
23145	A						
32415	B	52341	B	31425	B	51342	B
23451	C	53214	A	13452	C	53124	A
24315	A	35241	C	14325	A	35142	C
42351	C	32514	A	41352	C	31524	A
43215	A	23541	C	43125	A	13542	C
34251	C	25314	A	34152	C	15324	A
43521	B	52134	B	43512	B	51234	B
45312	A	25143	C	45321	A	15243	C
54312	D	21534	A	54321	D	12534	A
53421	A	12543	C	53412	A	21543	C
35412	C	15234	A	35421	C	25134	A
34521	A	51243	C	34512	A	52143	C
43251	B	15423	B	43152	B	25413	B
34215	C	14532	A	34125	C	24531	A
32451	A	41523	C	31452	A	42513	C
23415	C	45132	A	13425	C	45231	A
24351	A	54123	C	14352	A	54213	C
42315	C	51432	A	41325	C	52431	A
24135	B	15342	B	14235	B	52431	B
21453	A	51324	C	12453	A	52314	C
12435	C	53142	A	21435	C	53241	A
14253	A	35124	C	24153	A	35214	C
41235	C	31542	A	42135	C	32541	A
42153	A	13524	C	41253	A	23514	C
24513	B	31254	B	14523	B	32154	B
42531	C	32145	A	41532	C	31245	A
45213	A	23154	C	45123	A	13254	C
54231	C	21345	A	54132	C	12345	A
52413	A	12354	C	51423	A		
25431	C	13245	A	15432	C		

complete peal. By tradition, the method begins in the middle of a set, so the decomposition is carried out from the change 32415 instead of 12345. The box on p. 217 shows the results.

There are precisely two singles: both use move D. The permutations between the two Ds are the *even* permutations, which interchange an even number of pairs of bells: they are said to form the alternating group A_5. Without the two singles only half of the possible changes could be rung. Stedman's Doubles obeys all the rules of change-ringing and has a beautiful and intricate group-theoretic structure. You might like to experiment with a graphical representation.

"That's really quite remarkable," said Jane. "Those old bell-ringers must have known an awful lot of group theory!"

"Ze earliest work of any consequence on permutations was done by Joseph-Louis Lagrange in 1770 and Paolo Ruffini in 1799," said Quasimodulo. "Zey was studying ze solution of algebraic equations. Permutation groups is not invented until about 1815, when Augustin-Louis Cauchy, 'e write a 'uge paper on ze topic. 'Oo invented Stedman's Doubles?"

"Fabian Stedman explained it in his *Tintinnalogia*."

"Whe' did 'e write *zat*?"

"1668," said Jane.

ANSWERS

1. The value of $n!$ is approximately a googolplex when $n = 10^{98}$. A closer approximation is $n \sim 1.03 \times 10^{98}$.
2. F = FI = F(FLFLFL) = FFLFLFL = ILFLFL = LFLFL
3. The multiplication table for quick six is just like that for slow six, but with L and F interchanged:

1st string	I	F	LF	FLF	LFLF	FLFLF
2d string						
I	I	F	LF	FLF	LFLF	FLFLF
F	F	I	FLF	LF	FLFLF	LFLF
LF	LF	FLFLF	LFLF	F	I	FLF
FLF	FLF	LFLF	FLFLF	I	F	LF
LFLF	LFLF	FLF	I	FLFLF	LF	F
FLFLF	FLFLF	LF	F	LFLF	FLF	I

FURTHER READING

Budden, F. J. *The Fascination of Groups.* Cambridge: Cambridge University Press, 1972.

Fletcher, T. J. Campanological groups. *American Mathematical Monthly* 63 (1956): 619–26.

Rankin, R. A. A campanological problem in group theory. *Mathematical Proceedings of the Cambridge Philosophical Society* 44 (1948): 17–25.

Stewart, Ian. *Ah! Les Beaux Groupes.* Paris: Belin, 1982.

White, A. T. Ringing the cosets. *American Mathematical Monthly* 94 (1987) 721–46.

CHAPTER

14

A Six-Pack for the Tree-God

———

"...So I says to him, no, I *can't* fix you up with a nice little eclipse just in time for the spring fertility rite!" Rockchopper Rocknuttersson, the president of the Obeliskers' Guild, otherwise known as Rocky — Rocky II if you wanted to be formal — was in a bad mood. Bogpeople usually were. "Ruddy priests, always asking the impossible. If he wants a ruddy eclipse, I told him, why doesn't he sacrifice seventeen goats to the Moon-goddess himself? But no, seems it's the wrong time in the Sothic cycle, whatever that means. Mumbo-jumbo, I reckon."

"Typical," agreed his apprentice, a skinny lad too young to have earned a patronymic and thus simply named Pnerd. "Always the same. Every longbarrow has to be longer than the previous one, just to outdo the priests in the next village. I'll be glad when we evolve out of the neolithic age."

"But that wasn't all of it," continued Rocky. "Seems Archpriest Moloch Molochsson's come up with this great idea for placating the gods, and he want us to build it."

"Another stone circle?" asked Pnerd hopefully. The Obeliskers' Guild had long ago diversified into all kinds of masonry work, and a crafty apprentice could pick up plenty of overtime when a stone circle was going up.

"No. Circles are out, apparently. New directive from the acolytes."

"Oh," said Pnerd, disappointed. Then he perked up. "Stone avenue? Plenty of profit in a good, long stone avenue. Could do them an Avenue of the Thousand Idols—sounds good, eh? Easy to lay out, too; I mean, all you need is a few reels of pigskin twine and an accurate knotter. Plus an astrologer, of course, to make sure the Dog Star's lined up proper—"

"No, ley lines are out, too. Some new theory that the gods abhor a straight line."

"An altar," suggested Pnerd in desperation. "Can't have a good sacrifice without an altar. Nice dressed sandstone, looks hard but cuts soft, easy to chisel channels to carry the blood straight into the chalice, no fuss, no mess. You *can't* tell me they don't want an altar."

"Oh, they want an altar all right," said Rocky.

"That's all right then."

"Archpriest Moloch Molochsson wants *six* altars. One for each of the six taproots of Willo the Tree-god."

"I thought it was seven taproots."

"Look, the priests should know how many taproots a ruddy *god*'s got, right? That's their *job*, isn't it?"

"As you wish. Mmmm . . . it'd be worth setting up a production line for seven altars."

"Six."

"Right, six, yeah. 'Any rock as long as it's sandstone.' Great advertising slogan."

"It's not that altars that's the trouble! There's a planning regulation."

"Drains, I bet. It's *always* drains. Well, surely, a few gold obles to sweeten the Barrow Architect's Department and you can get them to drop any—"

"No, the priesthood has written a requirement into the tender. 'All distances between altars must be a whole number of paces.' Gods know why, something to do with Bountguzzle the Gluttony-goddess not liking any bits left over."

Pnerd thought about it. "Put 'em in a straight line," he said. "Ten paces apart. Then the distances will just be 10, 20, 30, 40, and 50 paces. Those are all whole numbers."

"Look, Pnerd, don't you ever listen? I just told you they didn't want any straight lines. No circles, neither. 'No more than two altars to be in a straight line and no more than three in a circle.'"

"Crikey, why settle for that?" asked Pnerd in surprise. "Never known the priesthood to compromise before. Why not ask for no more than one in a straight line and no more than one in a circle?"

"Beats me," said Rocky. "Some ritual restriction, I expect. Like not wearing three moleskins to the Offering of the Nine Virgins if it happens on Wongleday."

"Oh. So there's a good reason after all. Pity." They sat in gloomy silence. Pnerd put a clay pot on the fire and gnawed idly at a rabbit bone while he waited for the water to come to the boil. Nothing like a good pot of crabgrass tea when you had a problem.

They sipped in silence.

"I know," said Pnerd suddenly. "Snitchswisher Wishsnitchersdorter!" Rocknuttersson had made the customary gesture to ward off demons before he realized that Pnerd hadn't actually sneezed.

"Who's she?"

"A friend," said Pnerd guardedly. "Her dad's some kind of high palooka in the Society of Thaumaturgy. In fact, come to think of it, he *is* High Palooka. But," he continued hurriedly, noticing the look on Rocknuttersson's face, "that's not the point. Snitchswisher's a numerosophist. She's a wizard with numbers and things like that."

"A female cannot be a wizard," said Rocky.

"I was speaking metaphysically," Pnerd pointed out.

Rocky put down his pot of crabgrass tea. "She lives over at the edge of Dead Cat Swamp, doesn't she?"

"In it, actually. Hut on stilts, with a gangplank. Nice thatched roof."

Rocky shrugged. "Why not? Nothing much to lose."

"Except the contract for six sandstone altars," said Pnerd.

At their conference, Snitchswisher's immediate comment was: "Very interesting." Pnerd thought she looked really pretty in her stitched stoatskin wraparound trimmed with mouse ears. "You're looking for what we numerosophists call a 6-pack."

"No, if I want beer I–"

"If N is any number," said Snitchswisher, "then an N-pack is a set of N points in the plane, all lying at whole-number distances from each other, and such that no three lie in a straight line and no four in a circle." She wrinkled her nose in thought. "Do the priests want the altars to stand in a paved area?"

"No," said Rocky.

"Good. They always want square paving-stones, you see, because Hoph the Frog-demon has four heads, and they always want the altars placed over the centers of slabs, because they're afraid that if they don't do that, then the boglins that live under the slabs will escape through the cracks.

"Putting the altars on a paved area would introduce an extra condition into the problem: all the points would have to lie on the integer lattice [Figure 96]. That kind of arrangement is called an N-cluster. Every cluster is a pack, but not every pack is a cluster. Clusters are generally harder to find than packs, so we're lucky really."

PROBLEM ❶

Find a pack that is not a cluster.

Snitchswisher drew a picture (Figure 97) in the muddy ground with a stick. "Here's the smallest 3-cluster. It's just the usual 3-4-5–triangle that surveyors use to lay out right angles."

"Oh, yeah," said Pnerd. "I know about that one."

"It's an application of Pa Thuggerass's Theorem," said Snitchswisher.

"You mean old Thuggerbottom Thugstranglersson? Goes round all day with his head in the air, falling into ditches?"

"He's a very accomplished numerosophist," said Snitchswisher defensively. "He discovered a general rule for 3-clusters. And he realized that you can fit four copies of his 3-cluster together to make a 4-cluster." She drew Figure 98.

"Don't you mean 5-cluster?" asked Rocky.

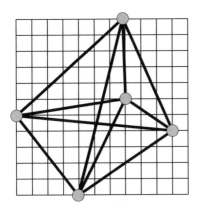

FIGURE **96** An *N*-cluster is an *N*-pack on the integer lattice.

"No, you have to miss out the middle point — shown as an open circle — or else you get three in a straight line. The same construction works with any Pa Thuggerean triangle, as we call them, like 5-12-13 or 8-15-17."

"Aha! And if you stick *three* of them together? Does *that* give a 5-cluster?"

"No, it's not that easy. You have to do it like this." She showed them the smallest 5-cluster: see Figure 99.

PROBLEM ❷

If the solid circle is the origin, what are the coordinates of the five points in Figure 99?

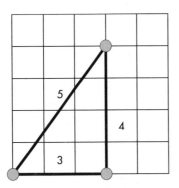

FIGURE **97** The smallest 3-cluster.

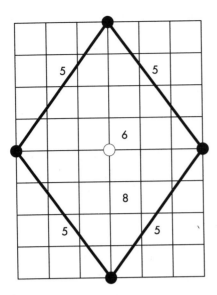

FIGURE **98**   Four 3-clusters make a 4-cluster. Not a 5-cluster, because the central point must be removed in order not to break the rule about three collinear points.

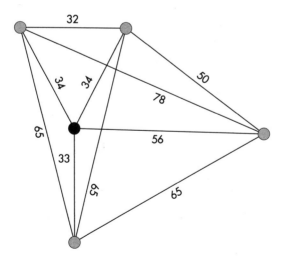

FIGURE **99**   The smallest 5-cluster. For the coordinates, see the answer to Problem 2.

"Great! So all we need is to find a 6-cluster, or even just a 6-pack. You know how to do that, of course."

Snitchswisher shook her head, and the shrew skulls set in her ears rattled entrancingly. "No, I don't."

"You're sure?"

"Absolutely. If only the condition about not lying on circles was relaxed, then I could find an N-cluster for arbitrarily large N [see the box below]. But the circle condition rules that method out. The answer, if there is one, lies in the distant future."

"Oh," said Pnerd. He thought about it for a moment. "Not much use to us, then, is it? I mean," he added, "if it was lying in the village cesspit, or out in the middle of Noisome Midden, or even in the Cave of the

NON-COLLINEAR LATTICE POINTS WITH ALL DISTANCES INTEGRAL

It is possible to find an arbitrarily large finite set of points, all lying on the same circle, having integer coordinates, and such that all distances between them are integers.

Let A be an angle in a 3-4-5 triangle (Figure 100A) so that cos $A = \frac{4}{5}$, sin $A = \frac{3}{5}$. In particular these are both rational numbers. Draw radial lines from the origin at angles 0, 2A, 4A, 6A, 8A, . . . 2($N-1$)A, meeting the unit circle at points P_0, P_1, P_2, . . . P_{N-1} (Figure 100B). Their coordinates are rational because sin $n\theta$ and cos $n\theta$ are polynomials in sin θ and cos θ with integer coefficients. The distances between points P_n and P_m are 2|sin ($n-m$) A| and by standard trigonometric formulas these are also rational. Now dilate by the lowest common denominator of all the rationals that occur to get points with integer coordinates and at integer distances.

By letting N become infinite, the same construction, without the final dilation, yields an infinite set of points, all lying on the same circle, having rational coordinates, and such that all distances between them are rational.

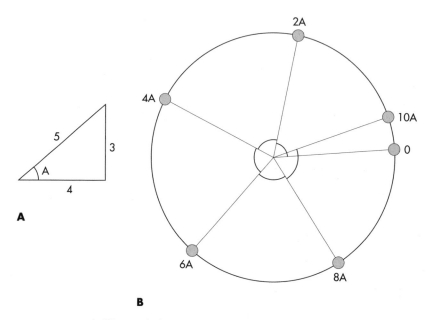

FIGURE 100   *A.* The angle in a 3-4-5 triangle. *B.* Constructing a set of *N* points, so that all distances are rational.

Dragonbreathed—er, Dragon, we could try to get it, um, couldn't we?" He withered beneath Snitchswisher's gaze. "Sorry, only trying to help. Well, thanks anyway. I guess we'll be going now. Give my regards to your dad."

They were halfway down the gangplank across the swamp when she called them back. "No, wait! My father! My father is a thaumaturge, he has ways to see into the future! Perhaps he can help!" She eyed Rocknuttersson frankly. "Though you'll have to offer him a cut of the profits, he's got a sound business head on his shoulders, you know, even though he pretends to be a senile old duffer."

Shortly she reappeared with her father, Wishsnitcher Dishpitchersson. "What you need," said Wishsnitcher, "is carrots."

"Tarot, father, not carrot," said Snitchswisher.

"Or potatoes," the old man continued, not missing a beat. "Very popular method, is spudomancy—divination by vegetables, y'know. But new potatoes are best and it's not the season. I'll use my Tarot cards instead."

He produced a pack from behind a pottery vase of bogwhort, riffling them with the practiced skill of a croupier, and dealt. "Hmmm, the Fool, and upside down to boot. Visibility's not so good today," he said. "I put it down to pollution from the flintworks. All those flakes. Mmmph, a bit of silver to grease the temporal lobes, that'd do it. *Un*fortunately I — you haven't, by any chance — oh, most generous. Hrrmph. I warn you I'll have to retain the coins after the ritual. Soak up the magic, they do, extremely dangerous to carry them around if you're not a trained thaumaturge. Fine, don't mention it, most generous indeed. Now, yes . . . ah, I sense a message coming through, from the year . . . the square-wheeled Chariot, the stoned Hermit . . . yes, 1983. The Dangling Man, the Tower . . . your problem will be answered by William Kalsow and Bryan Rosenburg; also independently by Landon Noll and David Bell. Snitchswisher, write this down. From a chosen point, pace 546 paces east, 272 paces north; also 155 paces east, 540 paces south . . ." He continued until he had given a complete description of the coordinates of points in the smallest 6-cluster (Figure 101).

"Great!" said Pnerd.

"I dunno . . . it's a bit *big*," said Rocky. "I'm not sure there's room on Sainsbury Plain for altars that far apart. Can't you find something smaller?"

"Not if you want a 6-cluster, lying on the integer lattice," said the old man. "This one is the smallest. But if you don't want it on the integer lattice, a 6-pack would do and it might be smaller."

"We don't. Give it a whirl. Oh, yes, more silver — naturally, don't mention it, generosity's not the word."

The old man shuffled the cards and dealt again. "The Crescent Moon over the She-Devil, Temperance before Unnamed . . ." The last in hushed tones.

"Unnamed?"

"Pnerd, it's the *Taxman*, but nobody ever *says* that because it brings bad luck," whispered Snitchswisher.

"Oh. Sorry I —"

"Shhh! I've got to write this next lot down, it's complicated." And laboriously Snitchswisher took down the details of the 6-pack shown in Figure 102, which was proved to be the smallest possible 6-pack by Arnfried Kemnitz in 1989.

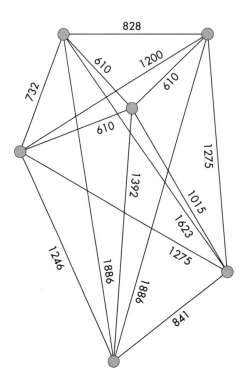

FIGURE 101   The smallest 6-cluster has points with coordinates (0,0),
(546,272), (132,720), (960, 720), (546,-1120), (1155,-540). It can be
contained in a circle of radius 1275.

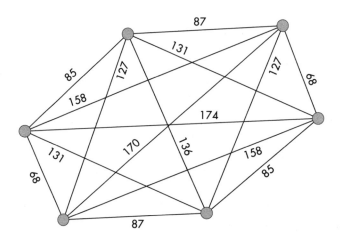

FIGURE 102   The smallest 6-pack is smaller than any 6-cluster.

"Is there any limit to the number of points in a cluster?" asked Pnerd recklessly. Rocknuttersson kicked him in the shin and handed over several more silver coins.

"None known."

"Oh. Maybe there might be an *infinite* cluster, then?" asked Pnerd, dodging Rocky's foot nimbly. But Snitchswisher Wishsnitchersdotter answered for her father. "No," she said. "That was proved long ago and is well known to numerosophists."

PROBLEM ❸

Prove that there is no infinite cluster. WARNING: This one's quite hard.

"That's a relief," said Rocky. "At least the priests can't ask us to build them an infinite number of altars."

"How about a 7-cluster?" said Pnerd. Rocky sighed but didn't even wiggle a toe.

"None known," said Wishsnitcher Dishpitchersson, at length.

"Impossible?" inquired his daughter.

"That's not known either. The problem's wide open. *Lots* of 6-clusters are known, though: Noll and Bell will list 91 of them in 1989, the largest being

$$(0,0) \quad (12852,8736) \quad (-7480,-15015)$$

$$(-4256, -17433) \quad (-17776,2457) \quad (0,-18753)$$

They do prove that no 7-cluster exists that fits inside a circle of radius 20,936, so if there is one, it has to be pretty big."

"Fortunately," said Rocky, "we don't need to know that. A 6-cluster will keep the priests happy. I shall go now and tender for the construction of six altars to symbolize the six taproots of Willo the Tree-god."

"The *six* taproots of Willo the Tree-god?" repeated Snitchswisher. "Is *that* why it has to be *six* altars?"

"That's what the priest said."

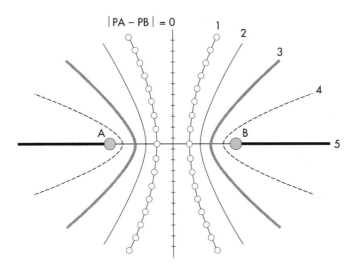

FIGURE 103   If PA and PB are both integers, then P lies on one of a finite set of hyperbolas.

"Funny," said Snitchswisher. "Oh, probably doesn't matter. They must have forgotten. Be all right as long as they don't check up. Mind you, Archpriest Moloch Molochsson's a real stickler for details."

"Check up?" asked Rocky.

"On what?" added Pnerd.

"Well, actually, Willo the Tree-god has seven taproots."

There was a long pause. "I *told* you—" began Pnerd. Rocky glared at him. "Nothing smaller than 20,936, eh?" Pnerd went on, trying to change the subject.

Rockchopper Rocknuttersson, president of the Obeliskers' Guild, held his head in his hands and howled.

ANSWERS

1. See Figure 102.

3. The coordinates are (0,0), (16,30),(−16,30), (0,−33), (56,0)

3. We show that all infinite sets of lattice points with all distances integral are collinear. Suppose we have a set of lattice points that contains three points A, B, C that are not collinear. Let k be the larger of the distances AB, BC. We claim that there are at most

$4(k + 1)^2$ points P such that PA-PB and PB-PC are both integers; this implies that the entire set of points must be finite.

We now prove the claim. By the triangle inequality, $|PA-PB| <$ AB, and thus assumes one of the values $0, 1, 2, \ldots, k$. Therefore P lies on one of $k + 1$ different hyperbolas with A and B as foci (Figure 103). Similarly, it lies on one of $k + 1$ different hyperbolas with B and C as foci. These families of hyperbolas meet in at most $(2k + 2)^2 = 4(k + 1)^2$ points.

FURTHER READING

Guy, Richard K. *Unsolved Problems in Number Theory.* New York: Springer-Verlag, 1981.

Hadwiger, Hugo, Hans Debrunner, and Victor Klee. *Combinatorial Geometry in the Plane.* New York: Holt, Rinehart and Winston, 1964.

Noll, Landon, and David Bell. *n*-clusters for $1 < n < 7$. *Mathematics of Computation* 53 (1989): 439–44.

Sheng, T. K. Rational polygons. *Journal of the Australian Mathematical Society* 6 (1966): 452–59.

CHAPTER 15

The Well-Tempered Calculator

"O blessed silence!" said Oliver Gurney.

"Don't be such a fussy old killjoy," said Deirdre. "*I* thought the 'Pheasant-Plucking Song' was rather good."

"I suppose," said Oliver grumpily. "But it doesn't suit the atmosphere of the Potted Dormouse. I knew that new landlord was going to cause trouble."

Oliver Gurney is eccentric, rotund, and he invents things—most of which lead to disaster. The *Potted Dormouse* is an extremely old pub east of Manchester: from the outside it's a gray stone box with a sign that closely resembles a pig in a fur coat. Oliver discovered it seven years ago while he was inventing a bacterium that converts oil into treacle, and he acts as if he owns it. Inside, for six days of the week, venerable Lancastrians eye each other stonily across their beer mugs and play shove-ha'penny, just as they have always done. But after the new landlord took over, Friday night is music night, and a bunch of local lads and lasses sing and play the guitar for a few hours.

"Anyway," Oliver went on, "that guitarist keeps getting his fingers all mixed up. At the very—"

"He only did it once," I interrupted. "He's got rather big fingers and he has trouble on the high notes when the frets get close together."

"Then he should get a guitar with the frets spaced farther apart."

"I don't think that would work," said Deirdre.

"No, it wouldn't," I confirmed. "There's a very good reason why the frets have to be spaced the way they are."

"To make the notes sound right, I suppose," said Deirdre, and I nodded. "But I don't see why the spacing gets smaller as the notes get higher," she added.

"Elementary physics of vibrating strings," said Oliver pompously. At the same time I said "Mathematics." Probably equally pompously, but I'm a poor judge of my own failings so I can't tell you for sure.

"I'm not very keen on mathematics *or* physics," said Deirdre. "Too impersonal. I prefer human things, like history and the arts."

"What fascinates me about music," I said, "is that it combines the lot: science, arts, culture, history . . . in fact, music is one of the oldest sciences. It was music, more than anything else, that led the Pythagoreans to believe that the universe is a harmonious place governed by numbers."

"Music a *science?*" said Deirdre in astonishment. Oliver perked up at once: he *loves* anything scientific, but there are huge gaps in his knowledge. So for the next hour or so I gave them a guided tour of music as a mathematical endeavor.

Today's Western music is based upon a scale of notes, generally referred to by the letters A–G, together with symbols ♯ (sharp) and ♭ (flat). Starting from C, for example, successive notes are

   C♯ D♯   F♯ G♯ A♯

  C D EF G A   B

   D♭ E♭   G♭ A♭ B♭

and then it all repeats with C, but one octave higher. On a piano the white keys are C D E F G A B, and the black keys are the sharps and flats. This is a very curious system: some notes seem to have two names, while others, such as B#, are not represented at all.

Of course there's more to it than that, and appearances are rather deceptive.

Today's system evolved over a long period of time, and it's a compromise between conflicting requirements, all of which trace back to the Pythagorean cult of ancient Greece. For convenience, I'll use the modern notation when giving examples, but purists will rightly object that I'm confusing slightly different ideas.

Claudius Ptolemy, who flourished around A.D. 150, is best known for his astronomical and geographical work, but he also wrote a book on the theory of music, called the *Harmonics*. Here Ptolemy reports the Pythagorean contention that the intervals between musical notes can be represented by whole number ratios. They demonstrated this experimentally using a rather clumsy device known as a canon (Figure 104*A*), a sort of one-string guitar.

If you slide the movable bridge along a canon, certain positions seem to produce notes that are more harmonious than others when compared with the note sounded by a full string. The most basic such interval is the octave: on a piano it is a gap of eight white notes. On a canon, it is the interval between the note played by a full string (Figure 104*A*) and that played by one of exactly half the length (Figure 104*B*). Thus the ratio of the length of string that produces a given note, to the length that produces its octave, is 2/1. This is true independently of the pitch of the original note. Other whole-number ratios produce harmonious intervals as well. The main ones are the fourth, a ratio of 4/3 (Figure 104*C*) and the fifth, a ratio of 3/2 (Figure 104*D*). Starting at a base note of C these are

```
C  D E  F   G A B   C
|       |  |        |
base    fourth fifth    octave
```

and you can probably see where the names came from. Other intervals are formed by combining these building blocks.

It is thought that, in order to create a harmonious scale, the Pythagoreans began at a base note and ascended in fifths. This yields a series of notes played by strings whose lengths have the ratios

$$1 \quad \left(\frac{3}{2}\right) \quad \left(\frac{3}{2}\right)^2 \quad \left(\frac{3}{2}\right)^3 \quad \left(\frac{3}{2}\right)^4 \quad \left(\frac{3}{2}\right)^5$$

or

$$1 \qquad \frac{3}{2} \qquad \frac{9}{4} \qquad \frac{27}{8} \qquad \frac{81}{16} \qquad \frac{243}{32}$$

Most of these notes lie outside a single octave, that is, the ratios are greater than 2/1. But we can descend from them in octaves (dividing successively by 2) until the ratios lie between 1/1 and 2/1. Then we rearrange the ratios in numerical order, to get

$$1 \qquad \frac{9}{8} \qquad \frac{81}{64} \qquad \frac{3}{2} \qquad \frac{27}{16} \qquad \frac{23}{128}$$

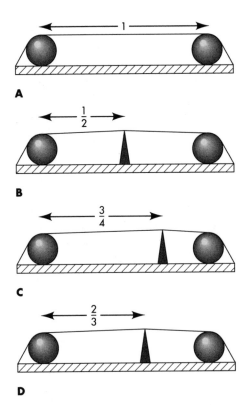

A

B

C

D

FIGURE 104   The canon, an experimental device used by the ancient Greeks to study musical ratios. *A.* Full string sounds base note. *B.* String 1/2 the length (ratio 2/1) sounds note an octave above base note. *C.* String 3/4 the length (ratio 4/3) sounds note a fourth above base note. *D.* String 2/3 the length (ratio 3/2) sounds note a fifth above base note.

On a piano, these correspond approximately to the notes

C D E G A B

and, as the notation suggests, there's something missing! The gap between 81/64 and 3/2 sounds "bigger" than the others. We can plug the gap neatly by adding in the fourth, a ratio of 4/3, which is F on the piano. In fact, we could have incorporated it from the start if we had *descended* from the base note by a fifth, adding the ratio 2/3 to the front of the sequence, and then ascended by an octave to get $2 \times (2/3) = (4/3)$.

The resulting scale corresponds approximately to the white notes on the piano, shown in Figure 105. The last line shows the intervals between successive notes, also expressed as ratios. There are exactly two different ratios: the tone 9/8 and the semitone 256/243.

It is here that the black notes of the piano, the sharps and flats, come in. An interval of two semitones is $(256/243)^2$, or 65,536/59,049, which is approximately 1.11. A tone is a ratio of $9/8 = 1.125$. These are not quite the same, but nevertheless it looks as if two semitones make a tone. This means that there are gaps in the scale: each tone interval must be divided up into two intervals, each being as close as is feasible to a semitone.

There are various schemes for doing this. The so-called chromatic scale starts with the fractions $(3/2)^n$ for $n = -6, -5, \ldots, 5, 6$. It reduces them to the same octave by repeatedly multiplying or dividing by two, and then places them in order. The result is shown in Figure 106. Each sharp bears a ratio 2,187/2,048 to the note below it, and from which it takes its name; each flat bears a ratio 2,048/2,187 to the note

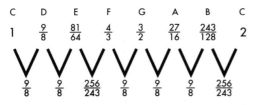

FIGURE 105   Scale formed purely from fifths and octaves approximates the white notes on a piano.

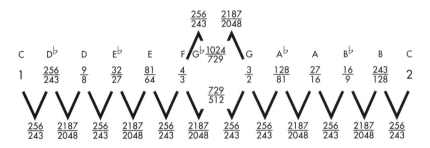

FIGURE 106   Chromatic scale of twelve notes, incorporating the black notes (sharps ♯ and flats ♭) as well. Two notes, F♯ and G♭, are trying to occupy the same slot.

above. There's a glitch in the middle: two notes, F♯ and G♭, are trying to occupy the same slot, but they differ very slightly from each other. There are many other schemes, also leading to distinctions between sharps and flats, but they all involve a 12-note scale that is very close to that formed by the white *and* black notes of the piano.

"Ah," said Oliver. "There's a good physical reason for all this, you know."

"You mean the waveforms—"

"Look, Ian, you've been talking for half an hour non-stop. It's my turn!" I apologized and he took up the tale.

"Y'see, Deirdre, when a string vibrates it does so as a standing wave [Figure 107]. And you have to fit a whole number of wavelengths in between the two ends, so that's where the Pythagorean whole numbers come in. When you sound a note on a guitar you don't just get a single wave along the string: you get harmonics with two waves, three, four, and so on. They all combine to give a richer sound.

"Now, if you combine two waves of slightly different wavelength, you get beats where they reinforce each other [Figure 108]. Those sound rather unpleasant to the ear."

"I think it may have something to do with the non-linear response of the eardrum," I put in. "There's probably a physiological rea—"

"The same problem occurs if *harmonics* of notes beat. The simplest way to avoid that is to use notes whose wavelengths are related by simple numerical ratios, say 3/2 or 4/3. So that's where the Pythagorean ratios come from, too."

"That makes sense," said Deirdre.

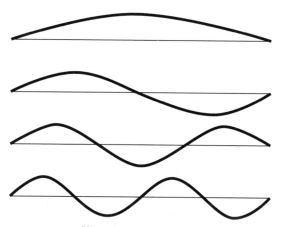

FIGURE **107**   Vibrations of a string form a standing sinusoidal wave. The length of the string is an integer number of wavelengths.

"Yes," I said, "but I still think you have to consider the physiol—"

"A good test of that theory," said Oliver, kicking me under the table, "was performed by Hermann von Helmholtz in 1877. He studied beats between harmonics, and used them to predict how the degree of dissonance between two notes should vary with the ratio between them. It agrees very well with psychological judgments made by human volunteers" (Figure 109).

"I *said* you have to consider the way an actual human ear—"

"I'll have another glass of wine, please, Oliver," said Deirdre tactfully, and sent him off to get one while I still had some shin left. That gave me time to pick up the story again.

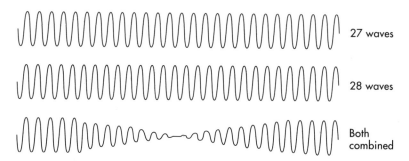

FIGURE **108**   Combining waves of slightly different wavelengths leads to unpleasant beats.

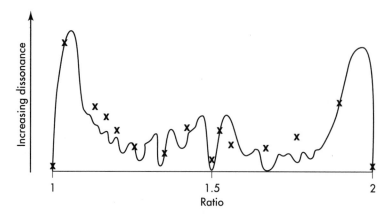

FIGURE 109   Helmholtz's theoretical curve of dissonance against musical interval (solid curve) compared with assessments by human observers (crosses).

As I explained to Deirdre, the reason for the glitch in the chromatic scale, and the reason that there are many different schemes for building scales, is that no "perfect" 12-note scale can be based on the Pythagorean ratios of 3/2 and 4/3. By a perfect scale I mean one where the ratios are

$$1 \quad r \quad r^2 \quad r^3 \quad r^4 \ldots r^{12} = 2$$

for a fixed number r. The Pythagorean ratios involve only the primes 2 and 3: every ratio is of the form $2^a 3^b$ for various integers a and b. For instance $243/128 = 2^{-7}3^5$. Suppose that $r = 2^a 3^b$ and $r^{12} = 2$. Then $2^{12a}3^{12b} = 2$, so $2^{12a-1} = 3^{-12b}$. But an integer power of 2 cannot equal an integer power of 3, by uniqueness of prime factorization.

PROBLEM ❶

Does it make any difference to this argument if the scale has a number of notes that differ from 12? Or if we allow other primes to occur in the ratios?

This property of primes puts paid to a musical scale based on Pythagorean principles of the harmony of whole numbers; but it doesn't mean we can't find a suitable number r. The equation $r^{12} = 2$ has a perfectly good solution, namely

$$r = \sqrt[12]{2} = 1.059463094. . .$$

The resulting scale is said to be equitempered.

If you start playing a Pythagorean scale somewhere in the middle — in other words; if you change key — then the sequence of intervals changes slightly. Equitempered scales don't have this problem, so they are useful if you want to play the same instrument in different keys. Musical instruments that must play fixed intervals, such as pianos and guitars, generally use the equitempered scale. The Pythagorean semitone interval is $256/243 = 1.05349. . .$, which is close to $\sqrt[12]{2}$, so the name semitone is used for the basic interval of the equitempered scale.

Deirdre thought about that for a moment. "You said that for vibrating strings, the musical interval is given by the ratio of the lengths. So how does that lead to the positions of the frets on a guitar?"

"Well," I said, "think about the first fret along, corresponding to an increase in pitch of one semitone. The length of string that is allowed to vibrate has to be $1/r$ times the length of the complete string. So the distance to the first fret is $1 - 1/r$ times the length of the complete string. To get the next distance, you just observe that everything has shrunk by a factor of r, so the spaces between successive frets are in the proportions

$$1 \qquad 1/r \qquad 1/r^2 \qquad 1/r^3$$

and so on. Now r is bigger than 1, so $1/r$ is less than 1, and that means that the distances between successive frets are *smaller*" (Figure 110).

Oliver returned with a dry white wine, plus two pints of Fosdick's Best Bitter ("The beer that refreshes parts you don't even have") and three bags of tripe-and-onion flavored crisps, which he adores and Deirdre and I detest. He always buys them because then he gets to eat the lot. By now he was his usual jovial self, and he launched into an animated explanation of how embarrassing it must have been for the Pythagoreans to discover that their beautiful numerical schemes had practical flaws.

I pointed out that when the Greeks were faced with irrational numbers such as $\sqrt[12]{2}$, which cannot be written as exact fractions, they

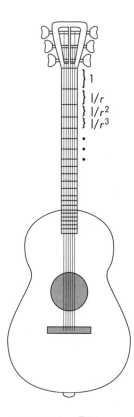

FIGURE 110   Distances between guitar frets shrink for the higher notes.

usually resorted to geometry. According to tradition, Greek geometry placed considerable emphasis on those lengths that can be constructed using only a ruler and a pair of compasses. For example, squares and square roots can be so constructed (Figure 111).

PROBLEM ❷

The ancient problem of "duplicating the cube" asks for such a construction for $\sqrt[3]{2}$. We now know that no such construction exists. (We also suspect that the emphasis on ruler and compasses was less strong than many history books try to claim, but that's another matter.) Deduce that there is no ruler-and-compass construction for $\sqrt[12]{2}$ either.

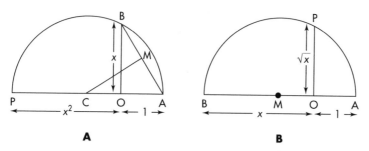

A   **B**

FIGURE 111   Constructing squares and square roots with ruler and compasses, given a line of unit length. *A*. Squares: Draw a right triangle AOB with OA = 1, OB = x. Find the midpoint M of AB and draw MC perpendicular to AB to meet the extension of AO at C. Draw a semicircle with center C through A, to meet the extension of AO at P. Then OP has length x^2. *B*. Square roots: Draw a line AOB with OA = 1, OB = x. Find the midpoint M of AB and draw the semicircle center M through B and A. Draw OP perpendicular to AB to cut the semicircle at P. Then OP has length $\sqrt{x}$.

"Yes, but," Deirdre pointed out, "you've said that the equitempered scale is a compromise, an approximation. And the true fourth, at an interval of 4/3, does in fact sound more harmonious than the equitempered fourth. Singers find it more natural, for example."

There was a stunned silence. "You know about this," I said.

"Yes, I studied musical theory at Huddersfield Poly. But you were explaining it all so nicely, I hated to interrupt."

Oliver started laughing and then stopped again when he realized she must also have heard about von Helmholtz. Deirdre picked up the thread of her remarks. "What I wanted to say was, since the equitempered scale is just a compromise, isn't there some *approximate* geometric construction that tells you where to put the frets on a guitar?"

That set me off again. You see, not only is there just such an approximate construction, but it has a very curious history indeed. The story illustrates the deep elegance of mathematics, but it is also a humbling tale: an outstanding triumph of a practical man nullified by a professional mathematician's carelessness.

"Oooh goody!" said Deirdre, and Oliver's eyes lit up—though that could have been the beer. "*Do* tell."

In the sixteenth and seventeenth centuries, finding geometrical methods for placing frets upon musical instruments—lute and viol rather than guitar—was a serious practical question. In 1581 Vincenzo Galilei, the father of the great Galileo Galilei, advocated the approximation

$$18/17 = 1.05882. . .$$

This led to a perfectly practical method that was in common use for several centuries. In 1636 Marin Mersenne, a monk better known for his prime numbers of the form 2^p-1, approximated an interval of four semitones by the ratio

$$\frac{2}{3 - \sqrt{2}}$$

Taking square roots twice, he could then obtain a better approximation to the interval for one semitone:

$$\sqrt{\sqrt{\left(\frac{2}{3 - \sqrt{2}}\right)}} = 1.05973. . .$$

which is certainly close enough for practical purposes. The formula involves only square roots and thus can be constructed geometrically. However, it is difficult to implement this construction in practice, because errors tend to build up. Something more accurate than Galilei's approximation, but easier to use than Mersenne's, was needed.

In 1743 Daniel Strähle, a craftsman with no mathematical training, published an article in the *Proceedings of the Swedish Academy* presenting a simple and practical construction (Figure 112). You might like to try it out, and compare with measurements from an actual instrument. But how accurate is it? The geometer and economist Jacob Faggot performed a trigonometric calculation to find out, and appended it to Strähle's article, concluding that the maximum error is 1.7%. This is about five times more than a musician would consider acceptable.

Faggot was a founding member of the Swedish Academy, served for three years as its secretary, and published eighteen articles in its *Proceedings.* In 1776 he was ranked as number four in the Academy: Carl Linnaeus, the botanist who set up the basic principles for classifying animals and plants into families and genera, then ranked just ahead of him in second place. So when Faggot declared that Strähle's method was

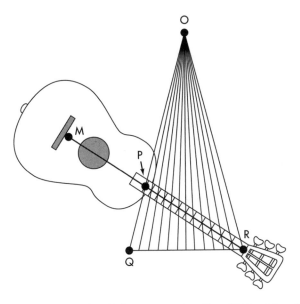

FIGURE 112   Strähle's construction. Let QR be 12 units long, divided into 12 equal intervals of length 1. Find O such that OQ = OR = 24. Join O to the equally spaced points along QR. Let P lie on OQ with PQ 7 units long. Draw RP and extend it to M so that PM = RP. If RM is the fundamental pitch and PM its octave, then the points of intersection of RP with the 11 successive rays from O are successive semitones within the octave, that is, the positions of the 11 frets between R and M.

insufficiently accurate, that was that. For example, F. W. Marpurg's *Treatise on Musical Temperament* of 1776 lists Faggot's conclusion without describing Strähle's method.

It was not until 1957 that J. M. Barbour of Michigan State University discovered that Faggot had made a mistake.

Faggot began by finding the base angle ∠OPQ of the main triangle: it is 75°31′. From this he could find the length RP and the angle ∠PRQ. Each of the 11 angles formed at the top of the main triangle by the rays from the base could also be calculated without difficulty: it was then simple enough to find the lengths cut off along the line RPM.

However, Faggot had computed ∠PRQ as 49°14′, when in fact it is 33°32′. This error, as Barbour puts it, "was fatal, since ∠PRQ was used in the solution of each of the other triangles, and exerted its baleful influence impartially upon them all." The mistake was equivalent to making PQ equal to 8.6 instead of 7. The maximum error reduces from 1.7% to 0.15%, which is perfectly acceptable.

Thus far the story puts mathematicians, if not mathematics itself, in something of a bad light. If only Faggot had bothered to *measure* $\angle$ PRQ! But Barbour went further, asking *why* Strähle's method is so accurate; and what he found is a beautiful illustration of the ability of mathematics to lay bare the reasons behind apparent coincidences. (I should say immediately that there is no suggestion that Strähle himself adopted a similar line of reasoning: as far as anyone knows his method was based upon the intuition of the craftsman rather than any specific mathematical principles. We shall see that his intuition was extremely good!)

The spacing of the nth fret along the line MPR can be represented on a graph (Figure 113A). We take the x-axis of the graph to be the line QR in Figure 112, with Q at the origin and R at 1. We move MPR so that it forms the y-axis of the graph, with M at the origin, P at 1, and R at 2. The successive frets are placed along the y-axis at the points 1, r, $r^2, \ldots, r^{11}, r^{12} = 2$. (Note that this differs from the ratios $1/r$, $1/r^2$, $\ldots$ mentioned above, because we are working from the opposite end of the string.)

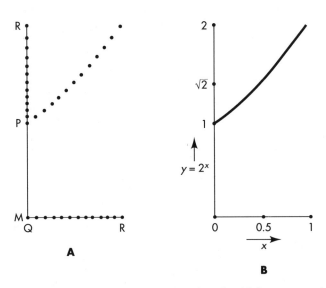

FIGURE 113   *A.* Graph representing Strähle's construction as a function. *B.* Finding the best-fitting fractional linear function by fitting it to the points $x = 0, \frac{1}{2}, 1$.

A mathematician would call Strähle's construction a projection with center O from a set of equally spaced points along QR to the desired points along MPR. It can be shown by simple geometric arguments that such a projection always has the algebraic form $y = (ax + b)/(cx + d)$ where a, b, c, d are constants. This is called a fractional linear function. For Strähle's method, the constants are a = 10, b = 24, c = −7, d = 24: the projection takes a given point x on QR to the point $y = (10x + 24)/(−7x + 24)$ on MPR. I'll call this formula Strähle's function, but we must remember that Strähle himself didn't write it down — it's just an algebraic version of his geometric construction. However, it is the key to the problem.

If the construction were exact, we would have $y = 2^x$. Then the 13 equally spaced points $x = n/12$ on QR, where $n = 0,1,2,. . .,12$, would be transformed to the points $2^{n/12} = (2^{1/12})^n = r^n$ on MPR, as desired for exact equal temperament.

But it's *not* exact, even though Barbour's calculations show that it's very accurate. Why? The clue is to find the *best possible approximation* to 2^x, valid in the range $0 \le x \le 1$, and of the form $(ax + b)/(cx + d)$. One way to do this is to require the two expressions to agree when $x = 0, \frac{1}{2}$, and 1 (Figure 113*B*). That gives three equations to solve for a, b, c, d; namely

$$b/d = 1$$

$$(\tfrac{1}{2}a + b)/(\tfrac{1}{2}c + d) = \sqrt{2}$$

$$(a + b)/(c + d) = 2$$

At first sight we seem to need one more equation to find four unknowns, but really we only need the *ratios* b/a, c/a, and d/a, so three equations are enough.

This approach leads to the values

$$a = 2 - \sqrt{2}$$

$$b = \sqrt{2}$$

$$c = 1 - \sqrt{2}$$

$$d = \sqrt{2}$$

so that the best possible approximation to 2^x by a fractional linear function takes the form

$$y = \frac{(2 - \sqrt{2})x + \sqrt{2}}{(1 - \sqrt{2})x + \sqrt{2}} \qquad (1)$$

"That doesn't look much like Strähle's function," said Deirdre.

"No," I agreed. "But now comes a final bit of nifty footwork."

"You change $\sqrt{2}$ to some approximation," suggested Oliver.

"Well, I wasn't actually going to do *quite* that. What Barbour did was to estimate the *error* in terms of the approximation 58/41 to $\sqrt{2}$. And Isaac Schoenberg did the same when he wrote about the problem in 1982. You see, if you substitute 58/41 for $\sqrt{2}$ in (1) then you get $(24x + 58)/(-17x + 58)$, which is different from Strähle's function.

"But now you come to mention it, that does seem a more natural thing to try. I doubt it will work, but let's see." I grabbed a napkin, borrowed a pen from Oliver, and started scribbling. The tools of the mathematician's trade are pencil and paper: in consequence, no mathematician ever carries either, and they *always* have to borrow a pen and scribble on a napkin.

Light began to dawn.

There is a series of rational numbers that approximate $\sqrt{2}$. One way to get them is to start from the equation $p/q = \sqrt{2}$ and square to get $p^2 = 2q^2$. Because $\sqrt{2}$ is irrational, you can't find integers p and q that satisfy this equation (or, more accurately, because you can't find integers p and q that satisfy this equation, $\sqrt{2}$ must be irrational!). But you can come close by looking for integers p and q such that p^2 is close to $2q^2$. The best approximations are those for which the error is smallest; that is, solutions of the equation $p^2 = 2q^2 \pm 1$. For example, $3^2 = 2.2^2 + 1$, and $3/2 = 1.5$ is moderately close to $\sqrt{2}$. The next case is $7^2 = 2.5^2 - 1$, leading to $7/5 = 1.4$, which is closer. Next comes $17^2 = 2.12^2 + 1$, yielding the approximation $17/12 = 1.4166.\ldots$, closer still. You can go on forever, and there's a beautiful theory that leads into continued fractions and Pell's equation and things like that.

What my scribbles had revealed was this. Divide the numerator and denominator of formula (1) by 2 and rewrite it as the equivalent formula

$$y = \frac{x + \dfrac{1}{\sqrt{2}}(1 - x)}{\dfrac{x}{2} + \dfrac{1}{\sqrt{2}}(1 - x)} \qquad (2)$$

Then replace $\sqrt{2}$ by the approximation 17/12, so that $1/\sqrt{2}$ becomes 12/17. This gives

$$y = \frac{x + \dfrac{12}{17}(1 - x)}{\dfrac{x}{2} + \dfrac{12}{17}(1 - x)} \qquad (3)$$

Finally, this simplifies to give $y = (10x + 24)/(-7x + 24)$, which is *precisely* Strähle's formula!

So Strähle's construction is very accurate because it effectively combines *two* good approximations:

- The best fractional linear approximation to 2^x is formula (1) above.
- Strähle's function is obtained from formula (1) by replacing $\sqrt{2}$ by the excellent approximation 17/12.

The errors corresponding to the various approximations discussed above are compared in Figure 114. The biggest errors are Faggot's!

"So," I finished up, "thanks to the mathematico-historical detective work of Barbour, we now know not only that Strähle's method is extremely accurate: we also have a very good idea of *why* it's so accurate. It's related to basic ideas in approximation theory and in number theory."

Which left just one question unanswered — and I fear forever unanswerable. It was raised by Oliver Gurney after he had digested an hour of mathematics in addition to his three packets of tripe-and-onion crisps.

"That's fascinating," he said. "Absolutely remarkable. The expert confounded and the practical man vindicated after a mere 218 years. Let it not be said that there's no justice in this world! If I ever meet Strähle in the afterlife I'll tell him; I'm sure he'll be pleased to have his name

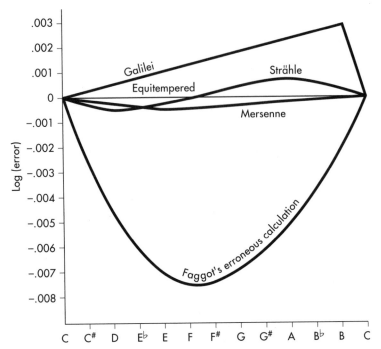

FIGURE 114   Errors in various constructions. The size of the error is measured by taking the logarithm of the ratio of the approximate value to the true value.

cleared. But what I'd really like to ask him is this: *How on earth did he think of his construction in the first place?*"

ANSWERS

1. No scale of finite length, with a constant ratio between notes that is a rational number, can ascend an exact octave—with the trivial exception of a scale that just goes up in octaves. Equivalently, the equation $r^n = 2$ can have no rational solution r when n is an integer greater than or equal to 2. To prove this, write r as a product of primes: $r = 2^a 3^b \ldots p^c$. Then $r^n = 2$ implies that $2^{na-1} 3^{nb} \ldots p^{nc} = 1$. By uniqueness of prime factorization, we must have $b = 0, \ldots, c = 0$. Also $na = 1$ so $n = 1$, $a = 1$, and the scale must go up in octaves.

2. If there were a ruler-and-compass construction for $^{12}\sqrt{2}$, then by squaring twice (using ruler and compasses as in Figure 111) we could construct $^{3}\sqrt{2}$, which we know is impossible. So there can be no ruler-and-compass construction for $^{12}\sqrt{2}$.

FURTHER READING

Barbour, J. M. A geometrical approximation to the roots of numbers. *American Mathematical Monthly* 64 (1957): 1–9.

Károlyi, Ottó. *Introducing Music.* Harmondsworth, England: Penguin Books, 1965.

Schoenberg, Isaac J. On the location of the frets on a guitar. *American Mathematical Monthly* 83 (1976): 550–52.

———. *Mathematical Time Exposures.* Washington, D.C.: Mathematical Association of America, 1982.

Stewart, Ian. *Galois Theory.* London and New York: Chapman and Hall, 1989.

CHAPTER 16

Sofa, So Good...

The Worm family had settled happily into their new bungahole, and Henry in particular was feeling unusually at peace with the world, having finally persuaded the builder to finish tiling the bathroom wall. "Henpecked" is a term that worms find unusually offensive, so we won't use it to describe Henry's customary state; but it was true that his wife, Anne-Lida, did have a very strong personality. Henry, left to his own devices, would have been perfectly happy spending his day with his tail up reading the sports pages of the *News of the Worm* in front of the television set.

But Henry was seldom left to his own devices.

Now, however, was an exception. Anne-Lida had gone shopping, taking baby Wermentrude with her to buy a new tight. He was curled up on his favorite sofa, the one they'd had for years, that had molded itself to him like a second skin, and everything was—

"*Henryyyyy!* Henry, where *are* you? Oh, *there* you are, skulking away as usual. Henry, there's a sale on at Wormbase, you remember, the store that builds furniture from your own designs."

"Yes, dear. Wonderful news," said Henry. He knew what was coming next.

"This would be an *excellent* opportunity to replace that dreadful old sofa!"

Henry had two choices. One was to agree to replace the sofa. The other was to have a blazing row and *then* agree to replace the sofa. Henry was not so much an optimist as an optimalist, and he recognized the solution that would optimize his peace of mind. "Of course. I'll order a new one at once. What color would you like?"

"Lime green and purple polka-dots," said Anne-Lida. "To match the orange and yellow striped carpet and the blue flock wallpaper with the pink daisies. But the most important thing, Henry, is not the color, but the *shape!*"

"What shape do you want, dear? I'd recommend something sofa-shaped, myself, but perhaps—"

"That, Henry, is what I can't decide! You see, I really want a big one—you know how little room there is to curl when mother visits!" (*Another cunning plan down the drain,* thought Henry.) "But you remember that terrible trouble I had with the removal men when they tried to get the grand piano into the laundry room!" Henry remembered it well: in the end they had knocked out one wall and dismantled the piano. Now the washing machine sat on top of the instrument and the tumble drier lay on its side beneath. They kept the soap powder in the piano stool. What he found totally impossible to remember was why Anne-Lida had wanted the piano in the laundry room in the first place. "What I want is a sofa big enough for us all to curl up on, but shaped so that it will be possible to get it down the hallway and into the curling room." (Worms, of course, don't have sitting rooms.)

At that point Henry's tendency to optimalize landed him in the biggest mess of his life. "Then, my dear, you should have it made in whatever shape will allow the *largest possible* sofa to be maneuvered into the curling room! The hallway has the same width everywhere: as I recall, the only real obstacle is that right-angled bend in the middle."

"A brilliant idea, Henry! And what might this largest shape *be?*"

"I have no idea, my pet, but I'm sure that one exists and I will find out for you what it is. . ."

Several days passed.

"Have you worked it out, yet, Henry? The sale will be over by the weekend!"

"Well . . . not *exactly* worked it out, Anne-Lida . . . I can solve the problem with the additional condition that the sofa is not rotated in any way. Taking the width of our hallway to be one unit, it's easy to see that the largest sofa that can be pushed round the corner without rotation is the unit square. It has to fit the width of the hall in both directions."

"That, Henry, is not a sofa! It is a table!"

"Yes, my little cabbage . . . but it did seem to me that if you allowed the sofa to turn through an angle, it ought to be possible to get something larger around the corner. My next idea was to use a longer but narrower rectangle and let it turn through a right angle as it goes round the corner. But that didn't lead to any improvement . . .

PROBLEM ❶

Why not? What size and shape is the largest rectangle that can be moved round the corner while being rotated through a right angle?

"My first substantial result was that the diameter of any connected sofa is bounded. I mean, my dear, that anything that is all in one piece and can move round the corner is smaller than some definite size: to be precise, its longest dimension can be no more than $2 + 2\sqrt{2} = 4.8284$ [Figure 115]. This immediately implies that its area is no more than that of a circle of equal diameter, namely $\pi(3 + 2\sqrt{2}) = 18.3105$; but on thinking harder I rapidly improved that to $2\sqrt{2} = 2.8248$."

PROBLEM ❷

Prove that the area of the largest sofa is no greater than $2\sqrt{2}$.

"Then I thought I had a breakthrough," said Henry with a trace of pride tinged with sadness. "It occurred to me that if the part of the sofa

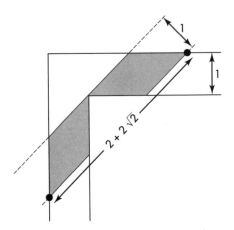

FIGURE 115   When the sofa is inclined at 45° it must be within the hallway, and also between two lines parallel to the dotted ones and 1 unit apart. The shaded region will disconnect if the slanting lines are lowered; so this position gives the largest possible diameter.

that scrapes against the corner is hollowed out, then it can be given a larger area." Henry sketched Figure 116. "I tried the general class of shapes like this: two quadrants of unit radius separated by a rectangle with a semicircular bite cut out of it. By choosing just the right length L for the rectangle, I more than doubled the area to 2.2074."

PROBLEM ❸

Among this class of shapes, what value of L *gives the largest area for a sofa that can get round the corner? What is that area exactly?*

"Mmmm," said Anne-Lida. "Not a bad shape for a sofa, Henry."
"No, my sweet. *But* I then realized that it is not optimal."
"Why not?"
"If you cut small pieces off the inside corners where the semicircle meets the straight bits, then the sofa fits more snugly round the sharp bit of corner, and you can fatten up the outside edges of the quadrants by more than you lose. The calculations get very messy, but I can certainly manage an area of 2.2164; and even that doesn't seem entirely optimal to me."

FIGURE 116   The Hammersley sofa. Add a quadrant of unit radius to each end of a $1 \times L$ rectangle and remove a semicircle. Which value of L gives the largest area, subject to being able to circumnavigate the corner?

"You mean you're stuck."

"In a word, yes."

"Then, Henry, you'd better get yourself *unstuck*! I want a new sofa and I want the largest one we can get! I absolutely *refuse* to settle for second best!"

"Yes, dear," said Henry. *I know, I'll do what I always do. . . I'll go and see Albert Wormstein,* he thought. Wormstein, who for some obscure reason worked in the Patent Office, seemed to know everything. *He'd* soon tell Henry the answer.

"You're in trouble," said Wormstein. "You've landed yourself with an old chestnut and it's a tough nut to crack. Nobody even knows where the question came from. Certainly John Horton Conway asked it in the '60s, but it's probably a lot older. At that time the object being moved was a piano, but in view of the obvious piano-sofa isomorphism I think we can conclude that the optimal piano must have the same shape as the optimal sofa. The first published reference that I know is by Leo Moser in 1966. The shape you found [Figure 116] was published soon after by J. M. Hammersley, as part of a tirade against 'Modern Mathematics,' and he conjectured that it is optimal. But at a meeting on convexity theory in Copenhagen (some say Ann Arbor) a group of seven mathematicians, including Conway, G. C. Shephard, and possibly Moser, did some informal work on the problem. In fact they worked on seven different variations — one each!" Two are shown in Figure 117; you might like to think about them for yourselves. "And they quickly proved that Hammersley's answer is *not* optimal, much as you did."

"Ah."

"Unfortunately, *they* couldn't find the best shape, either."

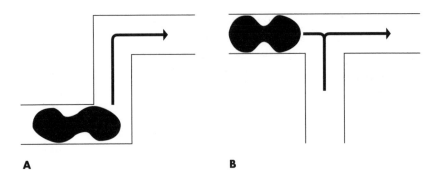

A   **B**

FIGURE 117   Two variations. *A.* The Shephard piano: get it around two bends. *B.* The Conway car: reverse it by turning into the side street.

"Oh."

"The problem has been open ever since, and nobody has yet published a solution."

"Oh, dear."

"Sorry, Henry old friend, but I can't help you."

Henry crawled sadly home to face Anne-Lida's wrath. But when he got there, she was wreathed in smiles. "Look at the lovely tight I bought Wermentrude! I've been hunting all week and finally found one in Woodworm's sale! Less than half price, too! The black-and-yellow zebra pattern brings out her slim build, don't you think? I bought six dozen. Oh, yes, and that Wormswine fellow phoned! He says he has important news for you about the sofa!"

Henry hastened to call Albert.

"Yes, Henry, marvelous news, hot off the press! I've just heard that Joseph Gerver, a mathematician at Rutgers University, has solved the Conway sofa problem!"

"Completely?" asked Henry, wary of Anne-Lida's wrath if some later improvement turned out to be possible.

"Well, there *are* a few gaps in his proof at the moment — if I were you, I wouldn't mention them to Anne-Lida. But they're all very plausible and the answer looks a good bet. In fact, Ben Logan of Bell Labs found the same shape in 1976 but didn't publish anything because his proof had the same gaps. Gerver discovered this when Andrew Odlyzko, Ronald Graham, and Jeff Lagarias all told him that his sofa looked very familiar, and Odlyzko tracked the work down later. So I'll call it the Gerver-Logan sofa."

"What shape is it?" asked Henry.

"Remarkable. Not dissimilar to Hammersley's attempt, but more subtle. The boundary comes in precisely eighteen separate pieces [Figure 118], each one being given by a single analytic formula! And the area is 2.2195, which improves on Hammersley's attempt by half a percent, and on your best effort by fourteen hundredths of a percent!"

"Do you mean to say that Anne-Lida's been giving me hell all week for a miserable fourteen hundredths of a percent?"

"Yes, Henry, but remember: precision is a fundamental principle of mathematics?"

Henry hurried round, and Albert filled him in on most of the details. I won't go into the more technical parts here, though I'll try to give you a feel for the way the argument goes, and what the gaps are. But first, let's see what happens as the Gerver-Logan sofa passes round the corner. We describe it in terms of the angle α through which the sofa has rotated. It begins with $\alpha = 0$ and by slides to the left until it is *almost* (but not quite) touching the outer left-hand wall. Then it pivots (Figure 119A), touching the right-hand wall at just two points and not touching any other walls at all, and tilts until α is just bigger than 2°. As this stage ends the sofa makes contact with the outer left wall and its innermost curve touches the sharp corner on the hallway; so that [Figure 119B] for α between 2° and about 39° it touches the walls at precisely four points. When α increases from 39° to 45° the lower right contact point lifts away, and the sofa swivels [Figure 119C] on just three contact points. The sequence from $\alpha = 45°$ to 90° is similar but in reverse because the sofa and hallway are symmetric.

This *describes* the solution, which is quite subtle! However, it does not explain where it comes from. Gerver has devised a complicated but

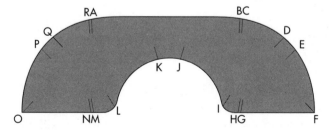

FIGURE 118   Gerver's solution (also found by Ben Logan at Bell Labs in 1976 but not published). The letters mark points on the boundary where the type of curve required changes.

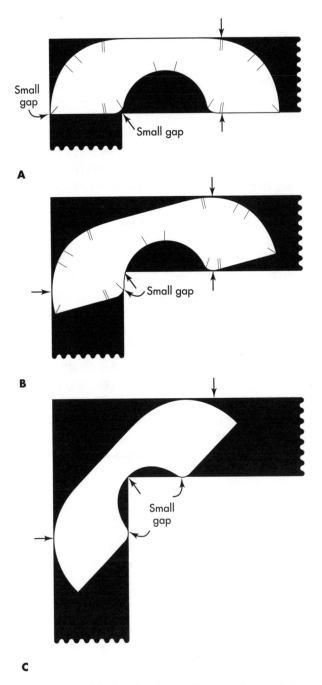

Small gap

Small gap

A

Small gap

B

Small gap

C

FIGURE **119** Moving the Gerver-Logan sofa round the corner. *A*. The first 2° of turn: there are only two points of contact (shown by arrows) with the walls. *B*. Between 2° and 39° there are four contact points. *C*. Between 39° and 45° there are three contact points. The sequence after 45° reverses these.

clear argument. It leads directly to the shape shown and the sequence of movements just described; but at present it contains a small number of unproved assumptions.

The first such assumption is that the shape rotates through a right angle as it passes round the corner. This *must* surely be true; but a proof seems surprisingly elusive. Whatever angle it rotates through, it must clearly fit into the hallway at all angles of inclination α. The first step is to prove the converse: any shape that fits the hallway at all angles of inclination α between 0° and 90° can in fact pass round it. This is *not* obvious, because the position at which it fits may not depend continuously on the angle. However, by inserting the sofa at angle α and then pushing it parallel to itself as far to the top and as far left as possible, continuity can be established.

Having settled that, we can for some purposes turn the problem on its head: fix the sofa and rotate the hallway! For each angle α between 0° and 90°, draw a copy of the hallway rotated to that angle. Call this the angle-α hallway. Then we want to maximize the area common to all these copies (their intersection) by sliding them without rotation—that is, by translating them. The next unproved assumption, also highly plausible, is that any optimal solution can be approximated as closely as we wish by the intersection of a finite set of these translated hallways. Such an approximate solution is a polygon with a large number of sides, resembling Figure 120.

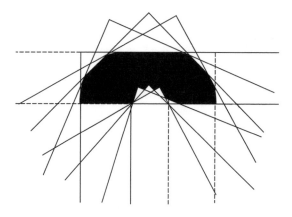

FIGURE 120   Polygonal approximation to an optimal sofa.

Such a polygon will touch the angle-α hallway along a number of sides. Suppose that for such sides the sum of the lengths that lie against the outer edge of the hallway is greater than the sum of the lengths that lie against the inner side. Then we can increase the area of the polygon by cutting off a thin slice on the inner side and adding one to the outer side, as in Figure 121. A similar argument holds if the sum of the lengths that lie against the outer edge of the hallway is smaller than the sum of the lengths that lie against the inner side. We conclude that in any optimal approximating polygon the two sums must always be equal, for any angle α.

In order to make a corresponding statement for the actual sofa, rather than a polygonal approximation, we pass to the limit in which the number of sides tends to infinity. To state the result we need the concept of the radius of curvature of a curve at a point, which is the radius of the best-approximating circle. A straight line has radius of curvature zero; the greater the amount of bend in a curve, the smaller its radius of curvature. Then the above result about approximating polygons leads, in the limit, to the following principle for the exact solution: for an optimal sofa, and for any angle α, the sum of the radii of curvature at points where the outer wall of the angle-α hallway touches the sofa must be equal to the corresponding sum for the inner wall.

This is an important constraint that must be satisfied by any optimal solution. For example we can now see that the Hammersley sofa (Figure 116) cannot possibly be optimal, because it fails to satisfy this constraint.

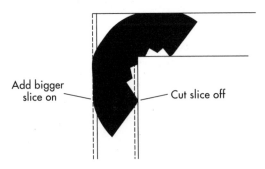

Add bigger slice on — — Cut slice off

FIGURE 121 If the outer edges are longer than the inner, then the area can be increased by adding a thin strip and removing a smaller one. Think of the entire hallway shifting slightly to the left (dotted lines).

Indeed, this is basically how Conway and his colleagues at Copenhagen realized that it could be improved upon.

The next step is to work out how the number of points of contact between the sofa and the walls changes as the angle α goes from 0° to 90°. Here there is a second gap in the proof: the scenario described in Figure 119 may not be the only one possible. However, Gerver has managed to rule out a great many variations, and nothing else looks likely. He therefore assumes that (as in Figure 119) there are two angles θ and ϕ with $0° < \phi < \theta < 45°$, at which the number of points of contact changes in the sequence 2, 4, 3; and that this sequence then reverses for angles greater than 45°. In Figure 119 these angles are ϕ ~ 2°, θ ~ 39°, but no values are assigned at the start of the proof. Gerver then defines three quantities $r(\alpha)$, $s(\alpha)$, and $u(\alpha)$, which are crucial to the whole analysis. The first two are the radius of curvature of the outer and inner boundaries of the sofa at the points for which the tangent makes angle α. The third quantity, $u(\alpha)$, is defined only for those angles at which the number of contact points is greater than 2, that is, between ϕ and 90°−ϕ. It is more complicated to describe, and the easiest way is by a picture (Figure 122). By using the above principle about the sums of radii of curvature, and more complicated statements of a similar kind, Gerver proves these equations:

$$\text{If } 0 < \alpha < \phi \text{ then } r(\alpha) = s(\alpha) = \tfrac{1}{2}.$$

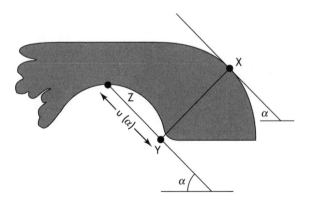

FIGURE 122   Definition of $u(\alpha)$. From a point X at which the tangent makes angle α, draw a perpendicular of length 1 to Y. Draw YZ at right angles to XY, hitting the edge at Z. Then $u(\alpha) = $ YZ.

If $\phi < \alpha < \theta$ then $r(\alpha) = s(\alpha) + u(90° - \alpha)$.

If $\theta < \alpha < 90° - \phi$ then $r(\alpha) = u(90° - \alpha)$.

If $\phi < \alpha < 90° - \phi$ and $\alpha \neq \theta$, $90° - \theta$, then
$u'(\alpha) = -u(90° - \alpha) - s(\alpha)$.

where $u'(\alpha)$ is the rate of change of u with respect to α.

He then solves these equations exactly using a mixture of calculus and geometry. The result is the shape you've already seen in Figure 118. Its area, as already mentioned, is approximately 2.2195. The two angles θ and ϕ turn out to be

$$\phi = 2.2448°$$

$$\theta = 39.0356°$$

The solution has left-right symmetry and is composed of 18 separate segments: nine on the outside of the sofa and nine corresponding ones on the inside. The arcs — other than straight lines — may appear circular in the picture, but most of them are not. However, they are well-known curves from classical differential geometry. To say what they are we need the concept of the involute of a curve: see Figure 123. Using this terminology and the labeling of Figure 118.

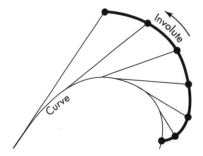

FIGURE 123   To obtain the involute of a curve, imagine a thin cord wrapped around it. As it unwraps, its free end traces out the involute.

AB is a straight line.

BC is the arc of a circle of radius $\frac{1}{2}$.

CD is the arc of an involute of a circle.

DE is the arc of an involute of a circle.

EF is the arc of the involute of the involute of a circle.

FG is a straight line.

GH is the arc of a circle of radius $\frac{1}{2}$.

HI is the arc of an involute of a circle.

IJ is the arc of an involute of an involute of a circle.

JK is the arc of an involute of a circle.

KL-RA are defined symmetrically.

Although there are a few gaps in Gerver's proof that the Gerver-Logan sofa is optimal, they concern only the general scenario that applies. Given that scenario (for example, the total angle turned through and the sequence of numbers of contact points) his solution is precise and unique. It is therefore without doubt at least "locally optimal" — the optimal solution that satisfies his scenario. And his sofa has larger area than anyone else's, so it is also the largest known sofa to date. But more is probably true. There are good reasons to believe that no other scenario is possible. Thus it is very likely indeed that Gerver and Logan have found the optimal sofa.

"Phew!" said Henry Worm as Albert finished. "If I'd known it was going to be as complicated as that when I started, I would have kept my stupid mouth shut!" Thanking Albert profusely, he rushed off to tell Anne-Lida the good news that her sofa could be fourteen hundredths of a percent bigger than she'd thought.

But when he got home, she was in a foul mood.

"Sofa? *Sofa*? Henry, I can't imagine why you think we need a new sofa! Our present sofa is *perfectly* adequate!"

"But, Anne-Lida, my pet . . . you wanted to buy a new sofa from Wormbase! I've spent all week—"

"Hmmph," she sniffed. "Wormbase indeed! I tell you, Henry, *there* is a shop that has lost *my* custom for good!"

"But — but *why*, dearest?"

"They refused to let me have lime green and purple polka-dot upholstery! Said they didn't stock it!" She snorted in anger. "Said they had no *intention* of stocking it! The salesworm was quite *rude* about it! Honestly, I don't know *what*. . ."

Henry crept silently away, back to the quiet comfort of his favorite, if non-optimal, sofa.

ANSWERS

1. The largest rectangle that can be moved around the corner while being rotated through a right angle has sides $\sqrt{2}$ and $1/\sqrt{2}$ (Figure 124). So its area is 1, the same as the unit square. (An easy way to see this is to note that the two white triangles have the same areas as the two light grey triangles; so the dark grey-and-white rectangle has the same area as the dark grey-and-light grey square.)

2. Figure 115 shows a strip of unit width rotated to an angle of 45°. I claim that the shaded area in that figure is the largest possible area for a sofa. To see this, consider the sofa as it goes round the corner, inclined at an angle of 45°. Then it must fit inside some strip parallel to the one shown in Figure 115 (and above it, or else the sofa is disconnected); say the strip between the heavy diagonal lines in Figure 125. It must also fit the hallway! Now, the area of the strips

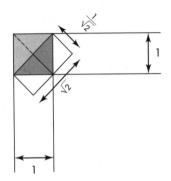

FIGURE **124** Largest rectangle that can turn round the corner.

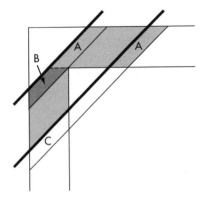

FIGURE **125**   Proof that the maximal area is no greater than $2\sqrt{2}$.

marked A are equal; but the area of strip B is *less* than that of strip C, unless the heavy lines coincide with the sloping lines in Figure 115, when the areas are equal. Therefore the shaded area between the heavy diagonal lines, inside which the sofa must lie, is less than or equal to that between the corresponding lines in Figure 115, whose area we already know is $2\sqrt{2}$.

3. The value of L in Figure 116 that gives the largest sofa of that particular shape that can pass round the corner is $L = 4/\pi$. The exact area is then $\frac{\pi}{2} + \frac{2}{\pi}$.

FURTHER READING

Gerver, Joseph L. On moving a sofa around a corner. *Geometriae Dedicata*. Forthcoming.

Hammersley, J. M. On the enfeeblement of mathematical skills by "Modern Mathematics" and similar soft intellectual trash in schools and universities. *Bulletin of the Institute for Mathematics and Its Applications* 4 (1968): 66–85.

Moser, L. Moving furniture through a hallway. *Society of Industrial and Applied Mathematics Review* 8 (1966): 381.

Wagner, N. R. The sofa problem. *American Mathematical Monthly* 83 (1976): 188–89.